# ECG

## IN THE POCKET

The Students' No-Nonsense Survival
Guide to Cracking ECGs

Copyright © 2025
Philipp Bergher

Detailed copyright and publishing information at the end of the book.

ISBN 979-8-89587-001-3

# Thank you! ...

**To Dr. Hannes Gänzer and Dr. Johannes Widkal for their meticulous proofreading, and to all test readers whose feedback greatly contributed to improving this book.**

—

I was in part inspired by the German book Elektro-Komiko-Graphie, published more than 50 years ago, which has helped countless readers grasp the fundamentals of electrocardiography in a uniquely approachable way.

Many thanks to the publisher Facultas for their careful copyright review of the manuscript and their kind permission to proceed.

# INTRODUCTION

Before we get started, let me give you a quick heads-up. It would be a shame to dive in completely unprepared.

This book was originally published in German, and now an adjusted international edition has been created. It is designed to give medical students a smooth introduction to the world of electrocardiography, tailored to their needs. That said, nurses and paramedics will certainly benefit from it as well. Many concepts are deliberately simplified, and it covers only the most essential basics of what can often be rather dry physiology. In other words, this book is aimed squarely at complete beginners and does not claim to meet the standards of specialist cardiology training.

The focus is on clinically relevant, practical examples. The overwhelming flood of information found in traditional ECG textbooks has no place here, especially since it tends to be daunting in the early stages. You'll find that this book contains everything you'll need for med school. Think of it as a springboard that helps you reach new heights, so you'll be better prepared to tackle more complex and comprehensive standard ECG books later on.

In many places, you'll spot a little stethoscope icon. This marks interesting clinical extras – the "nice to know" kind of facts. At the end of most chapters, you'll also find a compact summary containing everything you need to diagnose the pathology in question. This way, you can refresh your knowledge quickly whenever you need to.

**Note!**
The things you are about to learn here apply to fully grown adults. ECG interpretation in children differs in certain aspects, which can be explained by age-related anatomical and physiological differences.

With the solid, hands-on foundation you'll build here, deepening your ECG knowledge later on will be a whole lot easier. You'll learn how to find your way around an ECG and tackle interpretations step by step. By the end, you'll be able to confidently spot and diagnose the key basic pathologies.

**Enjoy the ride!**

# THE SYSTEM

The electrical impulses that stimulate the heart muscle, and make the heart beat, are generated, processed, and transmitted by specific anatomical structures within the heart. Together, they form the so-called cardiac conduction system.

To help you grasp these complex processes as intuitively and memorably as possible, this book uses a simplified visual model of the heart's electrical activity. But it still accounts for all the key physiological facts.

This fictional system is made up of several components that represent the real structures and, together, cover everything we need.

In this chapter, you'll get to know each component individually.

There's just one small prerequisite: you should already have a basic idea of the heart's anatomy and function. Since the focus of this pocket guide is deliberately on the ECG, the illustration below offers only an overview of the conduction system and doesn't dive into anatomical details.

You should at least be able to recognize basic terms such as atrium, ventricle, myocardium, and septum, and know what functional roles these structures play.

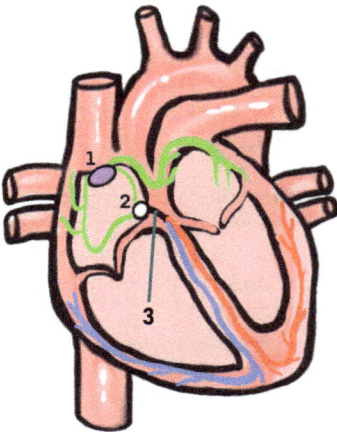

**1.** Sinoatrial (SA) node

**Green:** Conduction pathways of the atria

**2.** Atrioventricular (AV) node

**3.** Bundle of His

**Blue and Red:** Conduction pathways of the ventricles (bundle branches)

A small overview of the full, non-simplified anatomical cardiac conduction system and its components.

## The Power Plant

The sinoatrial (SA) node is like a power plant, generating the electrical impulses needed for the heart to contract entirely on its own, it's the starting point for everything. In resting mode, it produces such an impulse about 60–80 times per minute.

In addition, parts of our nervous system help regulate it and can influence the power plant. Under stress (sympathetic nervous system) or during relaxation (parasympathetic nervous system), adjustments happen automatically. So, the number of impulses per minute isn't always the same.

The SA power plant sits at the very top of the right atrium, marking the starting point of the heart's excitation. Along the entire conduction pathway, there are other "power plants," but the sinoatrial power plant is the strongest and, in a healthy heart, always sets the pace.

The many large chimneys make it clear that the sinoatrial power plant is our biggest and most powerful one.

## Nice to know – Automaticity

While the body's nervous system can influence the sinoatrial node, it doesn't actually set its rhythm. That job is done entirely by the SA node's specialized pacemaker cells. These cells have an intrinsic baseline activity that triggers regular discharges, making the heart beat completely independent of the body's nervous system.

This is why even in brain-dead patients, the heart can keep beating without interference. This ability is called cardiac automaticity.

**Think of it like this:** the SA power plant works much like a car engine idling at a set revolutions per minute on its own. The nervous system can only press the gas pedal and influence it indirectly, but it doesn't tell the engine how to run. The sympathetic nervous system pushes the pedal down, while the parasympathetic nervous system eases off the gas.

## The Conduction Pathways

Of course, we need a way to transport our electrical impulses, also called charges. In the real heart, they don't just spread randomly, but only through specific tissue.

This tissue is made up of specialized muscle cells arranged end to end, forming a kind of conduction pathway. Its job is to ensure the impulse travels in a precise direction and spreads quickly and evenly. This is called the cardiac conduction system.

In our simplified model, we'll imagine the parts of this pathway as pipes, but with one small twist:

**When an impulse rushes through one of these pipes, it creates a sound that we can "hear" and record.**

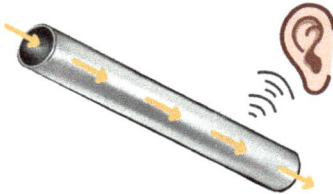

When a charge moves through a conduction pathway, you can easily visualize it by following the arrows.

## The Impulse

It is an electrical charge that travels along the conduction pathways. Guided by them, it races at high speed to even the most distant corners and recesses of the heart muscle.

**Two properties are worth noting:**

**1.** It may be very fast, but it still takes a certain amount of time to get from point A to point B. If you compare two conduction pathways of different lengths, you can measure a notable difference:

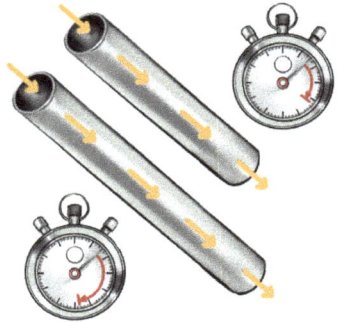

Longer distances take more time.

**2.** The thicker a conduction pathway is, the more electrical charge it can carry. The sinoatrial power plant feeds enough charge into the "pipe" to fill it completely.

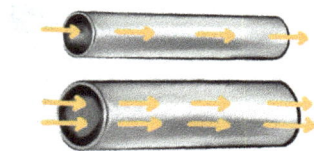

Thicker pipes automatically carry more charge.

## The Charging Station

The atrioventricular (AV) node is a structure located roughly in the middle of the conduction system, and it plays a crucial role. We'll imagine it as a charging station positioned right between the atria and the ventricles. It consists of a rotating axis with two gripping arms, each holding a battery.

A physiological impulse from the sinoatrial power plant in the atria can charge one of these batteries. Once it's full, the whole mechanism rotates and passes the battery's stored char-ge on to the ventricular conduction system.

The more the heart needs to hurry, the faster and more often new impulses from the sinoatrial power plant reach the station. The AV node adjusts to this pace, charging the batteries more quickly and passing them along faster.

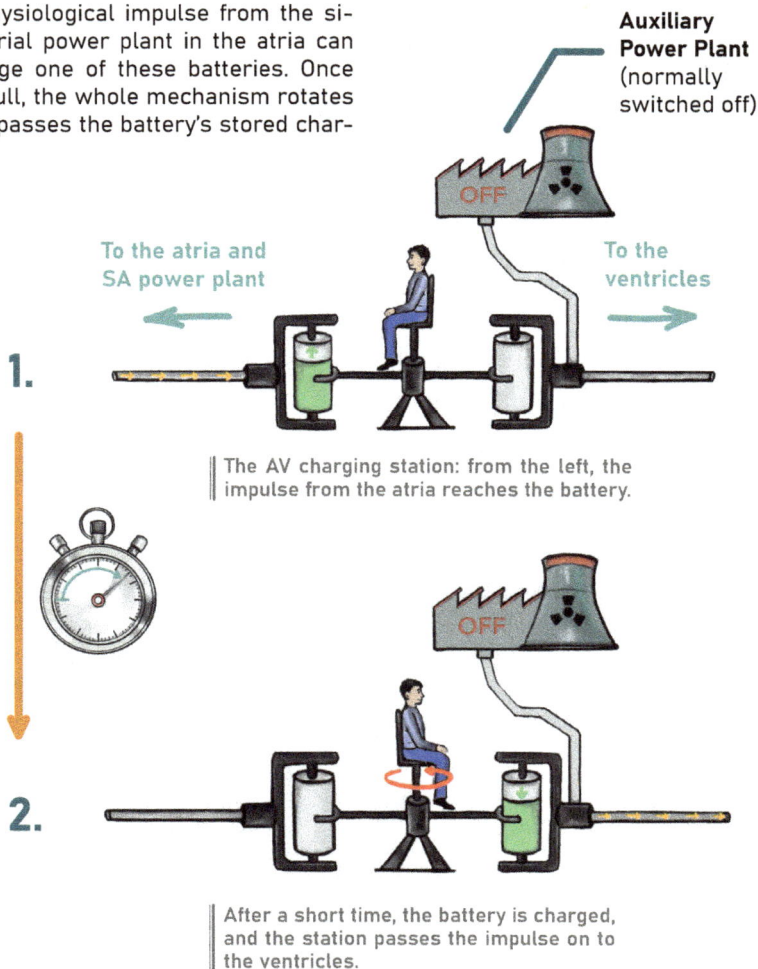

**Auxiliary Power Plant** (normally switched off)

To the atria and SA power plant

To the ventricles

**1.**

The AV charging station: from the left, the impulse from the atria reaches the battery.

**2.**

After a short time, the battery is charged, and the station passes the impulse on to the ventricles.

**Weak or faulty impulses can't fully charge the battery and therefore simply don't get passed on.** This is an important protective function, preventing potentially harmful rhythms, such as atrial fibrillation, from spreading to the ventricles.

**Charging a battery takes a little time, which slightly delays the transmission of the impulse.** This delay gives the atria time to empty their contents into the ventricles before the ventricles start building pressure. Without it, the entire heart would contract at the same time.

**The person "working" at the charging station also keeps an eye on whether any impulses from the sinoatrial power plant are coming in at all.** If none arrive, a small auxiliary power plant, built into the AV station and equipped with its own pacemaker cells, is quickly switched on. This plant sends impulses directly to the ventricles, but only at a maximum rate of about 40–50 times per minute.

> **Important!**
> If the AV station's auxiliary power plant also fails, there's another structure further down the line that can provide backup impulses - **the bundle of His.** However, it can only manage about 20 to 30 impulses per minute.
>
> The bundle of His is mentioned here because it's important for you to be aware of it. Since it isn't essential for understanding the pathologies discussed in this book, it won't be visually represented in our simplified model.

> **You should remember the three functions of the AV station:**
>
> 1. Filter out faulty impulses.
>
> 2. Delay ventricular activation.
>
> 3. Take over if the sinoatrial power plant fails.

## The Bundle Branches

They are simply conduction pathways that branch into two. For the atria, we can treat them as a single unit, so one conductor is enough to represent them. The ventricles, however, are different, because for many pathologies, it's important to distinguish between the left and right chambers.

In our model, we'll imagine two separate conductors located just beyond the AV station, each responsible for exciting one ventricle.

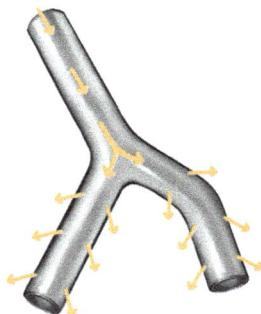

After splitting into the two bundle branches, the impulse spreads throughout the ventricles.

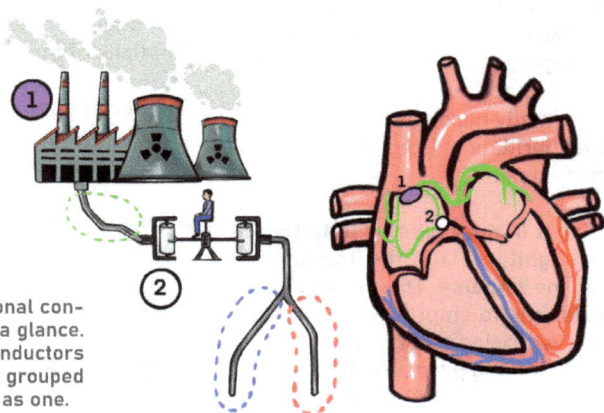

The complete fictional conduction system at a glance. The branched conductors of the atria can be grouped together and seen as one.

## It Gets Even Easier!

For many cases in this book, an even simpler illustration with short, straightforward pathways will do. To help you fully understand this perspective for later examples, the heart is shown here with dashed outlines.

As you can see, we're always looking at the heart from the front. This means the left ventricle is on the right side of the image, and the right ventricle is on the left. The atria are represented by

a single shared conductor into which the sinoatrial power plant sends its impulses. We don't actually need to include the power plant itself, since in most situations it's enough to picture the atrial conductor alone.

Next comes the AV charging station, represented by the square symbol in the center. Finally, we have the bundle branches.

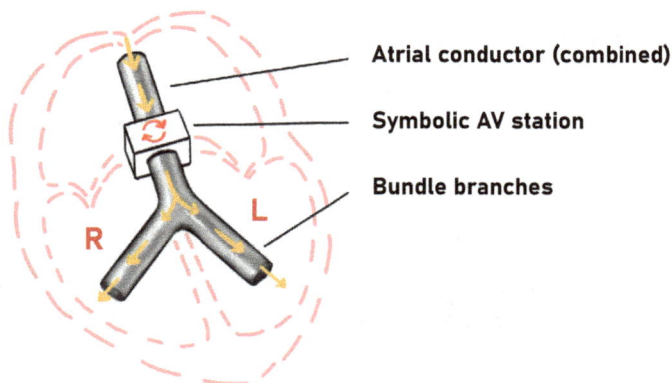

Atrial conductor (combined)

Symbolic AV station

Bundle branches

## The Leads

They form the interface between the heart's electrical activity and the jagged lines we see on the ECG.

For easier understanding, let's imagine that the leads act like little hearing devices. They "listen" to the sounds of tiny electrical charges rushing through our conductors during excitation, and they record them. Each lead is placed at its own fixed position on the body and "looks" at the heart from there. This means each one hears certain areas of the heart more loudly, while others sound quieter.

On the ECG, it works like this: when the charge moves toward one of our little hearing devices, the line for that lead moves upward. When the charge moves away, the line goes downward. The larger the amount of charge, the louder the "sound," and the bigger the spike on the paper, and vice versa.

In this book, the leads are represented in the simplest way, drawn as triangles labeled with their names. One corner of the triangle is always filled

**Important!**
Electrodes and leads are not the same thing!

Electrodes are simply the tools we use to create leads, they're not interchangeable terms. This distinction is especially important, because the first six leads are not located where the electrodes are actually attached to the body. You'll learn exactly how that works in the chapter "Navigation 101."

in with color. This shows the direction in which the lead is "looking."

In this example, they're labeled "A" and "B." These are two fictional leads that don't exist in real life. They follow no rules and can be placed anywhere we want. That's why they're used in some examples to make certain concepts easier to explain.

The real leads will be introduced to you in detail in one of the following chapters.

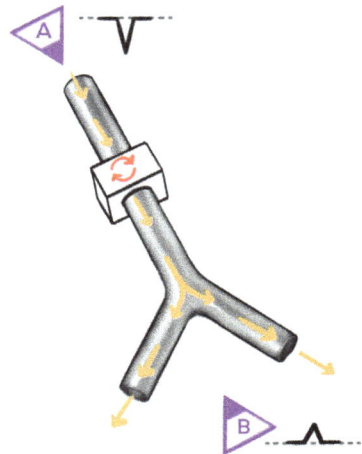

The impulse is moving away from lead A, so it records a negative deflection. Lead B detects the opposite, but since the charge isn't moving directly toward it, the spike is smaller in comparison.

# HOW THE HEART WORKS
## CARDIAC CYCLE

**Let's start right from the beginning: what exactly am I looking at in these spikes and waves?**

An ECG allows us to observe the electrical events that take place during the heart's contraction. Not just once, but many times in a row, since an ECG is usually recorded over several seconds, minutes or even longer.

What you're seeing here is one complete cycle of the heart, from the excitation of the atria to the recovery phase of the ventricles. **We call this the cardiac cycle:**

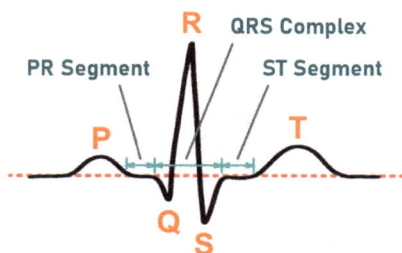

‖ The entire cardiac cycle at a glance.

**⚠ Note!**
Although they may look like jagged spikes or smooth waves on paper, in ECG terminology we always refer to them as waves.

To make it easier to work with, each wave is assigned a letter.

It's important to remember what they mean and what each section of the ECG represents. In this chapter, we'll break the cardiac cycle down into its individual components and go through them from left to right, explaining each one in detail.

## The Axes

The vertical axis (Y) on the ECG records the voltage of the electrical charge. It represents the strength of that charge. In our model, more charge means more strength. This would be the volume of the sound created by the moving impulse. The louder the sound, the higher the line rises.

The horizontal axis (X) records time - that is, how long a particular section of the ECG lasts.

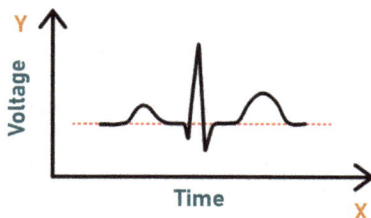

‖ The Axes of the ECG.

## P Wave

The P wave is produced when the impulse travels through the atria, causing them to contract and push blood into the ventricles. Changes in its shape or a prolonged duration can indicate possible problems with one or both atria.

The P wave represents the impulse from the "sinoatrial power plant." If it is consistently present and each ventricular activation is preceded by a P wave, we refer to this as a **sinus rhythm. It's the foundation of every normal, physiological ECG.**

## PR Segment

After the P wave, there is always a brief moment of "radio silence." We call this section the PR segment. **It extends from the end of the P wave to the beginning of the QRS complex.** At this point, the impulse is being delayed at our AV charging station, allowing the atria to empty completely before the ventricles start their work. That's why our leads pick up nothing here.

This section is particularly interesting, because if something is wrong with the AV charging station, we'll see it here.

> ⚠ **Note!**
> Despite what the name might suggest, the PR segment always ends at the very beginning of the QRS complex, it does not extend to the R wave.

## PR Interval

**This interval runs from the start of the P wave to the beginning of the QRS complex.** In other words, the PR interval includes both the P wave and the PR segment.

As you might guess, the PR interval can change not only with problems involving the AV station, but also if the P wave is unusually narrow or wide. That's why we use this value in ECG interpretation, it lets us catch two issues at once.

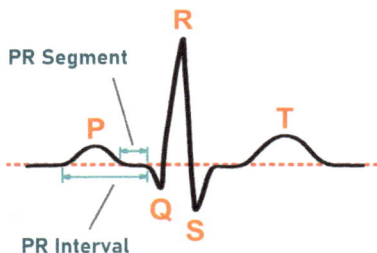

Risk of Confusion! PR Segment vs. PR Interval

> ⚠ **Note!**
> Never confuse the PR segment with the PR interval! The difference is small but important. A segment is always a flat (also called "isoelectric") portion of the cardiac cycle. There are two relevant ones: the PR segment and the ST segment.

## QRS Complex

The Q, R, and S waves together form the QRS complex, also called the ventricular complex. **It appears when the impulse passes through the ventricles, triggering them to contract.** During this time, the pressure inside the ventricles rises sharply. However, from a timing perspective, the blood is only pumped into the body after the QRS complex.

The appearance of the QRS complex varies depending on the lead. The skill lies in distinguishing a normal pattern from potentially pathological changes. The total duration of the complex is also assessed, as it can be prolonged in certain pathologies.

## ST Segment

After the QRS complex, there is once again a brief moment of "silence." The ventricles are fully excited (also called „depolarized"), with the electrical impulse having reached all parts of their muscle tissue, and nothing is "moving" within the conduction system. As a result, no sound is recorded. **This is already the early phase of repolarization. Meaning the recovery of the excitable cells.**

If something is wrong with the heart's blood supply, such as during a myocardial infarction, the ST segment often shows noticeable changes. We'll discuss exactly why that happens in the chapter on myocardial infarction.

The PR segment and the ST segment should normally be at the same level.

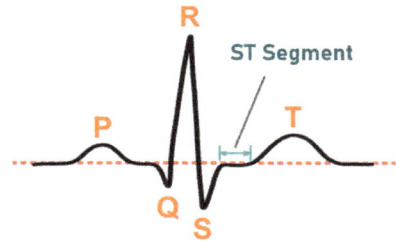

ST segment: Ventricular depolarization is complete, and for a brief moment nothing is happening.

## T Wave

**It represents the late phase of ventricular repolarization.** The muscle is still contracted and doesn't begin to relax until after the T wave, but the cells of the ventricular conduction system are already repolarizing and returning to their resting state so they're ready for the next excitation.

## Nice to know – Atrial Recovery

You might be wondering why, during the T wave, only the ventricular cells are recovering and not the atrial cells as well.

That's because the atria contract much earlier than the ventricles - they need to pump blood into them first. As a result, atrial repolarization also begins earlier than that of the ventricles. In other words, the atria have finished their job sooner and can start recovering earlier. The reason you can't see this phase on the ECG is that it occurs around the same time as the QRS complex and is simply hidden beneath it.

## QT Interval

This is the time span from the beginning of the QRS complex to the end of the T wave. It isn't relevant for most pathologies but can provide clues in certain cases. One example is a prolonged QT interval in hypocalcemia.

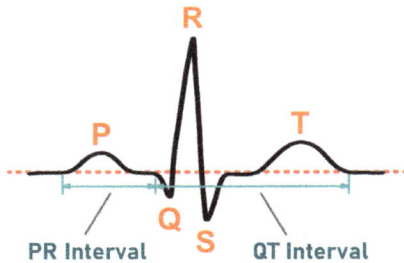

## The Isoelectric Line

The entire waveform always moves around a horizontal line that acts like a kind of foundation. Anything above this level is called positive, and anything below it is called negative. We call this the isoelectric line - it's shown as a dashed line in many of the illustrations in this book.

When nothing is happening, the ECG of a healthy heart always returns to this baseline, such as between the P wave and the QRS complex. In a real ECG, however, the line isn't actually drawn, we have to imagine where it runs. It is usually at the same level as the PR segment.

## Diastole and Systole

You may already be familiar with these two terms. They divide the heart's mechanical movements into two phases:

**Diastole** (filling) – After the previous beat, the heart relaxes and the ventricles begin to fill. The atria then contract, pushing even more fresh blood into the ventricles, where it waits to be pumped into the body.

**Systole** (ejection) – The ventricles start to contract. Once they are fully contracted, the valves open and blood is pumped into the body.

It gets interesting when we place both phases over the cardiac cycle and see how they line up. But don't worry, you don't need to memorize this. Still, it's helpful to visualize the sequence once and understand that the heart's electrical and mechanical activities occur slightly out of sync.

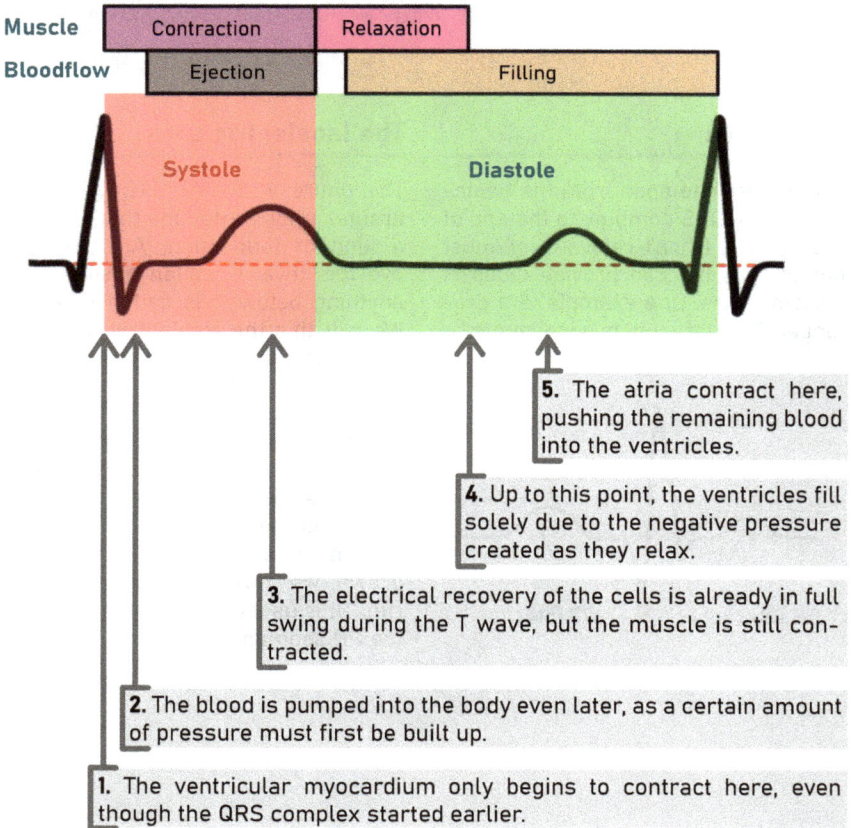

**5.** The atria contract here, pushing the remaining blood into the ventricles.

**4.** Up to this point, the ventricles fill solely due to the negative pressure created as they relax.

**3.** The electrical recovery of the cells is already in full swing during the T wave, but the muscle is still contracted.

**2.** The blood is pumped into the body even later, as a certain amount of pressure must first be built up.

**1.** The ventricular myocardium only begins to contract here, even though the QRS complex started earlier.

## The Labeling System

Now that you know the individual components of the cardiac cycle, let's take a closer look at the QRS complex. Its three components are named according to a specific system. When we later discuss pathologies that can be identified from the QRS complex, it's important for you to understand this system.

**Tip!**
Remember these two rules:

**1. The waves always appear in the same order (Q–R–S). If one is missing, the next takes its place.**

**2. Any positive wave is always labeled as an "R."**

With these rules, you can confidently label any ventricular complex.

**Important!**
A wave is only called a "Q" if it is negative and appears first in the complex – before any positive wave.

**The easiest way to explain this is with an example:**

The first wave in the figure below is positive. However, this doesn't mean it's a positive "Q" – it's an "R." According to the rules, any positive wave is always labeled as "R." That means there's no Q at all here, the R has taken its place.

Since the order is always the same, the negative wave following this R is labeled as "S."

The last wave is positive again, so we also call it "R." To tell the two apart, we write the first R in lowercase (since it's the smaller wave of the two) and add a prime mark to the second one.

This gives us an rSR' complex (pronounced simply "R-S-R").

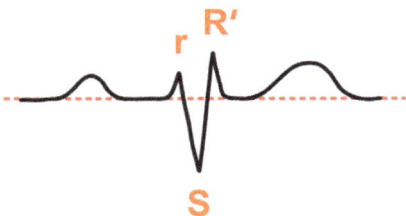

An rSR' complex is typically seen in a right bundle branch block, but more on that later.

# NAVIGATION 101

What comes next is meant to give your sense of direction a little boost and provide you with the essential knowledge you'll need so you won't get lost in the ECG jungle ever again.

You'll see how the electrodes you place on the body produce the different leads in the ECG, and what exactly the difference is between electrodes and leads.

## Where Do the Electrodes Go?

The setup is simple: we have three functional electrodes on the limbs. Those are red, yellow, and green. The fourth, black electrode serves only to suppress interfering signals. In addition, there are six more electrodes placed along the left side of the chest, extending to just below the armpit.

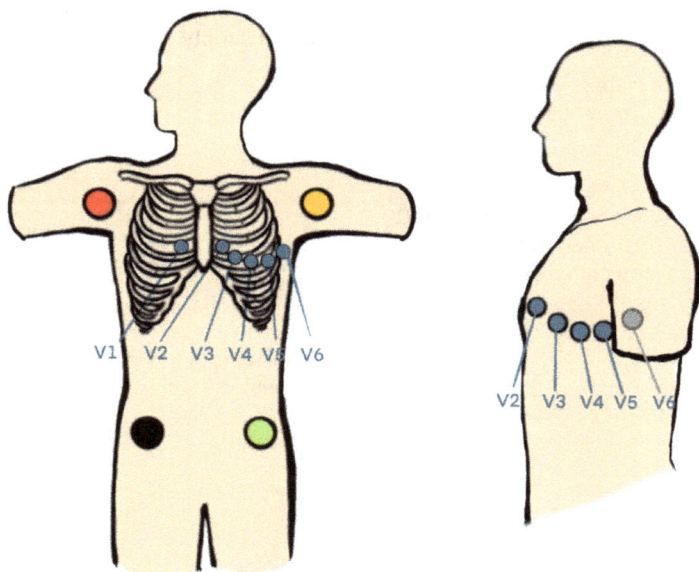

An overview of the individual electrode positions.

The best way to learn how to apply the ECG electrodes is to have someone show you a few times and then practice it yourself on a regular basis. Describing the entire process here in text form wouldn't be very practical.

## What Exactly Is a Lead?

And why can't you use the term "electrode" as a synonym? What's the difference?

A lead is essentially a directional reference. Each lead represents a specific angle from which the ECG "listens" to or "looks at" the heart, whichever way you prefer to picture it. You'll see both analogies used often throughout this book.

The electrodes, on the other hand, are merely tools for creating the leads. The raw signals they pick up aren't directly useful, they first need to be processed. The ECG computer automatically combines these signals in various ways by electrically linking the electrodes in different configurations. From these combinations, multiple "viewpoints," or leads, of the heart can be generated.

## Limb Leads

By combining them in this way, we are able to generate six different leads from the first three electrodes alone. These leads are arranged in a circle around the heart and "look" toward its center from their respective positions. They are called the limb leads and are designated as **I, II, III, aVF, aVR, and aVL.** You'll find an illustration on the next page.

## Precordial Leads

On the chest, we have another six leads available. Here, however, the number of leads is the same as the number of electrodes. Each electrode produces one lead.

All of them are also directed toward a central point in the heart. The chest leads are designated **V1 through V6.**

The term precordial comes from the fact that these leads are positioned directly in front of the heart (pre = in front, cor = heart).

## But Why Does This Matter?

Being able to generate these different "viewpoints" is crucial. Only when we give the ECG the ability to create spatial orientation can we determine from which direction or from which region a pathological signal originates.

**That's why we need to remember how each lead is oriented toward the heart and where it is located.**

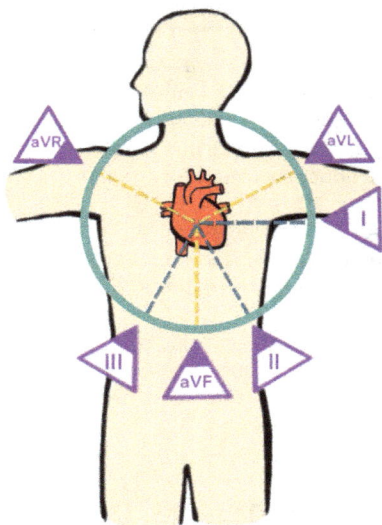

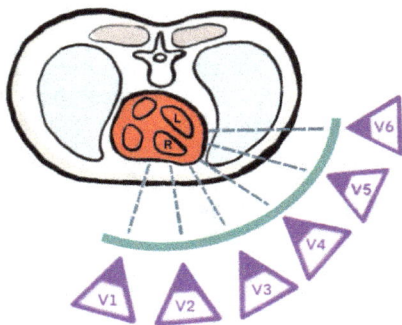

All the leads used in everyday clinical practice. Make sure to memorize their positions.

On the left are the limb leads, and on the right are the precordial leads.

## The Spatial Orientation of the Leads

As you may have already noticed, the **limb leads** are responsible for giving us orientation in the frontal plane. This means they mainly allow us to **distinguish between "up" and "down."**

aVR is directed from the top right, angled downward, and is therefore associated with the upper portions of the right ventricle. More importantly, however, aVL - together with Lead I - assesses the upper and lateral areas of the left ventricle.

Leads II, III, and aVF "look" at the heart from below, giving them a view of the inferior wall of the left ventricle.

The **precordial leads** give us access to the horizontal plane. They help us **distinguish between the left and right sides of the heart.** This allows us to detect pathologies affecting a specific ventricle or even the interventricular septum.

**Tip!**
Together, the leads work like a compass, showing us the direction of the pathology.

## How to Remember the Precordial Leads

We group the precordial leads in pairs according to their respective reference structures.

**V3 and V4** are the middle two of the six precordial leads. They "look" toward the central structure of the heart, roughly at the transition from the septum to the anterior wall. They primarily capture the anterior wall of the left ventricle.

**V5 and V6** are positioned to the left of V3/V4 and are therefore directed toward the left ventricle. They pick up all activity within the left bundle branch of the conduction system and have a clear "view" of the lateral wall of the heart.

**V1 and V2** lie to the right of the middle leads and are therefore directed toward the right ventricle. In addition to monitoring the activity of the right bundle branch, they focus primarily on the septum. **They can essentially "see through" the wall of the right ventricle to observe what is happening behind it.**

⚠️ **Important!**
We always look at the body from the front or from above! However, "left" and "right" always refer to the patient's perspective.

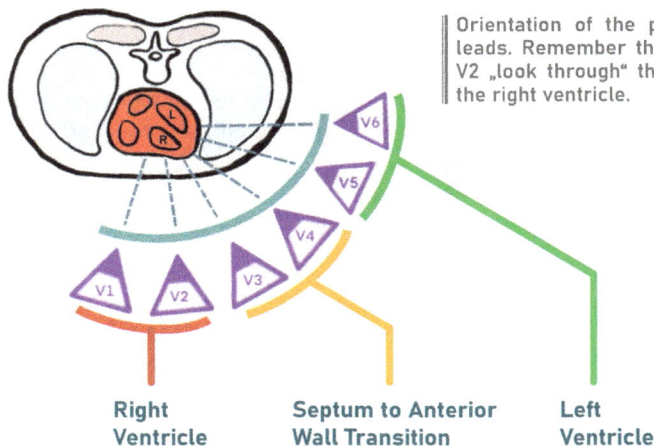

Orientation of the precordial leads. Remember that V1 and V2 „look through" the wall of the right ventricle.

**Right Ventricle**      **Septum to Anterior Wall Transition**      **Left Ventricle**

## Memorizing the Limb Lead Positions

The easiest way is to use certain landmarks on the body so you don't forget them.

• **Lead I** "listens" to the heart **from the left, horizontally.** It's the only lead that you simply have to memorize, as it has no landmark you can use for orientation.

• Moving clockwise, we arrive at the **left hip.** This is where **Lead II** is positioned.

• **Lead III** "looks" at the heart from the **right hip.**

The leads aVR, aVL, and aVF are even easier to remember. The last letter always indicates the location the lead "listens" from.

• **aVR** "listens" to the heart from the **right shoulder.**

• **aVL** from the **left shoulder.**

• **aVF** is the last of the limb leads and "observes" the heart **from the feet, straight up.**

You can now figure out which structures are being viewed by each lead by imagining how the leads "look" at the heart from their respective positions.

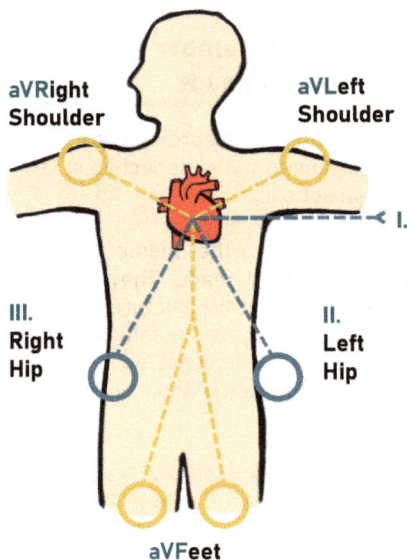

The landmarks of the limb leads.

**Important!**
Do I always need all 12 leads?

Usually, yes. However, if you only want to make very simple assessments - such as determining rhythm, heart rate, or atrial activity - then, in rare cases, just four electrodes on the limbs and thus the first six leads will suffice.

**By the way:** To make things easier to follow, the examples of pathologies in the following chapters won't always show the signals from all leads, but only those that are relevant to the specific case and display a characteristic change.

# EVERYTHING AT A GLANCE

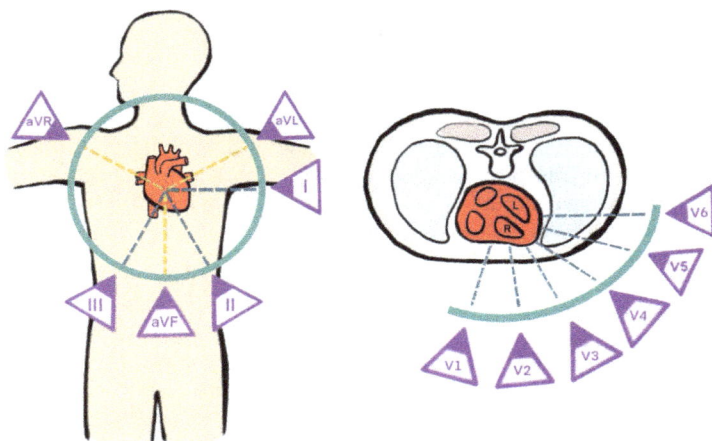

**Limb Leads** &gt; Distinguishing top from bottom (and left).

**Precordial Leads** &gt; Distinguishing left from right.

| aVR | right ventricular outflow tract |
|---|---|
| aVL, I | upper and lateral region of the left ventricle (lateral wall) |
| aVF, II, III | lower region of the left ventricle (inferior wall) |
| V1, V2 | right bundle branch and septum |
| V3, V4 | anterior wall of the left ventricle |
| V5, V6 | left bundle branch and left lateral wall |

# QRS AXIS MADE EASY

Before we talk about interpretation, we need to clear up two things you'll need for it. This chapter and the next focus on how impulses spread through the heart. We'll start with the so-called QRS axis.

## What Does It Mean?

Determining the QRS axis means figuring out the position of the heart's electrical axis in the frontal plane.

**In principle, it tells us in which direction the heart's impulses are spreading, when we project their course onto the body as seen from the front.**

Usually, it's not interpreted on its own, but together with other, more meaningful findings in the ECG, since by itself it only provides vague information.

As a student you should at least know how to determine it so that someone with more experience can make use of that information when needed.

**Tip!**
Picture it like this: If we were looking at a person from the front, their chest slightly transparent, and we could see the heart's impulses spreading with the naked eye, we could draw an arrow on their chest to indicate the general direction of that spread. That arrow would represent the electrical axis in the frontal plane.

Since people are usually not transparent, we use Leads I, II and III to determine it.

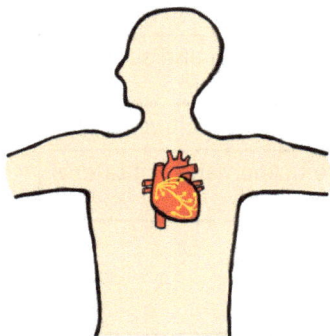

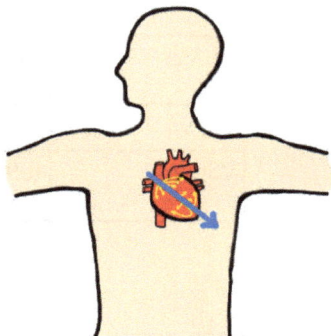

We can approximate the spread axis of the impulses with an arrow.

## Axis Types vs. Axis Deviation

Historically – and still common in German-speaking countries – the entire physiological range of axis orientation was divided into several small zones, called axis types. Depending on the axis orientation, a specific axis type was then assigned to the heart.

Internationally, this system is not used, as the clinical value of dividing the axis into multiple types has increasingly been called into question. Instead, a single normal range is defined, and the axis orientation is assessed to see whether it matches that range or deviates from it.

This approach is more clinically relevant and reflects current knowledge in ECG interpretation. It is also much easier to remember. That is why this book focuses on the international system.

Here, you'll learn a simple, visual technique that lets you assess the heart's axis at a glance.

## Here's How!

Picture a kind of compass with Leads I, II, and III arranged from left to right:

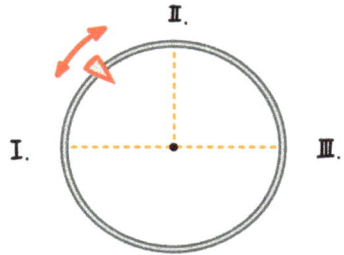

On a circular track around it is a movable red indicator. It works by being pulled toward or pushed away by the forces of the QRS complexes of the leads. The more positive a lead, the stronger its pull – the more negative, the stronger its push.

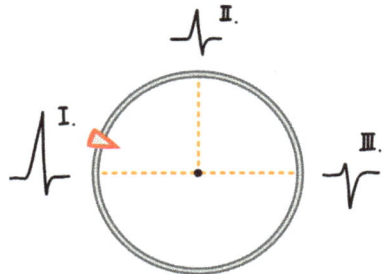

Lead I pulls the indicator mostly toward itself. Lead II acts slightly against it, as it is still somewhat positive. Lead III pushes it away. As a result, the indicator stays in this position.

For assessing the QRS axis, the upper half of the compass is what matters. As a visual aid, it is divided into two quarters with dashed lines.

## Practical Use of the Axis Compass

One ingredient is still missing before we are able work with the compass: we need to define the normal range.

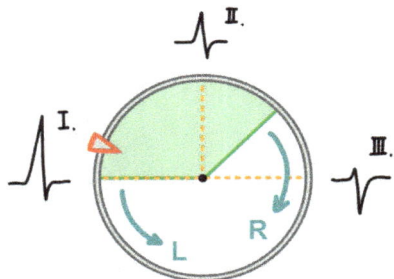

The normal range includes the entire left quarter as well as the first half of the right quarter.

If the indicator lies within the green area, the axis is normal. If it crosses the boundaries, we speak of a left or right axis deviation.

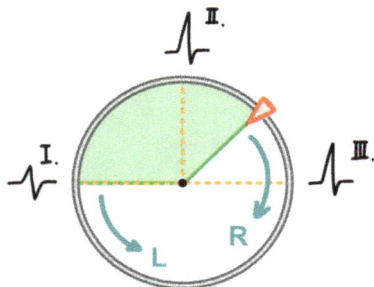

Leads II and III are equally positive. Lead I is neutral, as its positive and negative components are equal in size. The indicator therefore ends up in the middle of the right quarter, still just within the normal range.

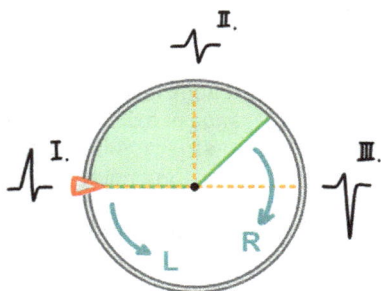

Lead I is the only one pulling, Lead II is neutral, and Lead III pushes away. The axis is far to the left, but still technically within the normal range.

## Crossing the Boundary

The previous examples have probably already helped you get a feel for the dynamics of the compass. But you may still be wondering how it's possible to cross the boundaries of the green area.

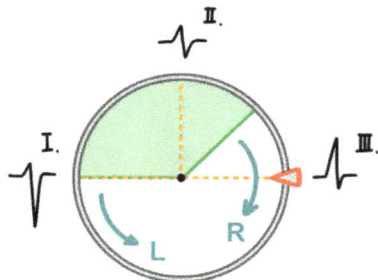

Right axis deviation: In this case, the QRS axis polarity is essentially the mirror image of the previous example. While the pattern from before falls within the normal range, its mirrored counterpart is already abnormal, because the normal range ends earlier on the right side.

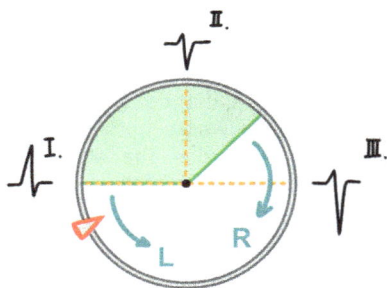

Left axis deviation: Only Lead I pulls, while Lead III pushes strongly away. Lead II also has a repelling effect, driving the indicator beyond the boundary.

With a bit of practice, you can picture the compass in your mind and quickly estimate where the QRS axis lies.

**Tip!**
If you have trouble visualizing things, you can instead just memorize the borderline cases. Leads I and III determine which side the axis is on, while Lead II decides whether there is a deviation:

**Left:** I strongly positive, III strongly negative
**Left axis deviation:** II negative

**Right:** I strongly negative, III strongly positive
**Right axis deviation:** II less positive than III or negative

Or you could also look directly at the ECG printout. Many machines display the QRS axis in degrees. In adults, the **normal range is roughly –30° to +90°.**

## Making Sense of Axis Deviations

Determining the QRS axis is important because it depends on several factors, most notably the spread of the heart's electrical impulses. This, in turn, is influenced to some extent by myocardial dimensions and the anatomical axis of the heart.

So, if something is abnormal in any of these areas, it can affect the axis.

Let's place the normal range of our compass over the body to better understand the connection:

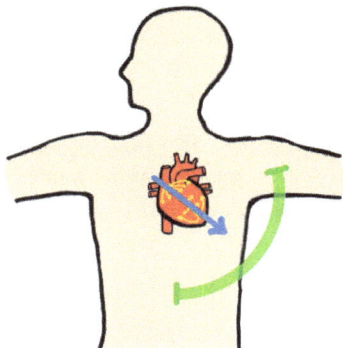

In a healthy heart like this one, we expect a QRS axis (blue arrow) that falls within the physiological normal range (green).

If we find that the QRS axis lies outside the normal range on an ECG, this indicates a pathological spread of impulses within the heart.

The causes can vary widely, from isolated severe hypertrophy of a ventricle to a pathology of the conduction system.

Fortunately, the QRS axis is usually not the only abnormal finding when a specific pathology is present. In most cases, there are additional, more specific clues in the ECG that help with the diagnosis.

**Important!**
Even though anatomical conditions can influence the QRS axis, that doesn't mean the relationship works both ways! We cannot determine the anatomical position of the heart based on the electrical axis. For that, an imaging technique is required.

## Nice to Know – Variability of the QRS Axis

Pregnant women often show a leftward shift of the axis, since the fetus pushes the abdominal organs upward, causing the heart to sit in a more horizontal position.

If you watch a live ECG recording, you can sometimes see how deep inhalation and exhalation change the axis, as the diaphragm moves up and down. This movement shifts the position of the heart, and with it, the direction of impulse propagation.

**Still, an ECG cannot provide reliable information about the anatomical position of the heart!**

# EVERYTHING AT A GLANCE

**Picture a compass with Leads I, II, and III arranged from left to right.** The QRS complexes influence the position of the red indicator, which shows the axis. The more positive the QRS complex of a lead, the stronger its pull on the indicator – the more negative, the stronger its repelling effect. If a complex is equally positive and negative, its forces cancel each other out.

The normal range includes the entire left quarter and half of the right quarter. If the indicator falls outside this range, a left or right axis deviation is present.

**For example:**

Normal QRS axis: Lead I pulls, Lead III pushes away, and Lead II adds a slight pulling effect.

Right axis deviation: Lead I pushes away, Lead III pulls, and Lead II is neutral.

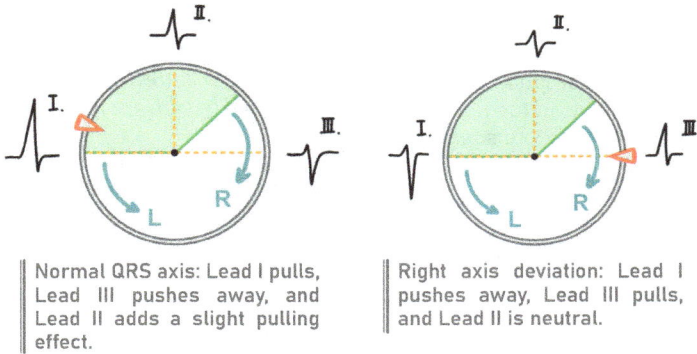

**Alternatively, you can memorize the left and right boundaries: Lead II then determines whether there is a deviation or not.**

**Left** – Lead I strongly positive, Lead III strongly negative
Left axis deviation if Lead II is also negative

**Right** – Lead I strongly negative, Lead III strongly positive
Right axis deviation if Lead II is less positive than Lead III, or negative

**QRS Axis normal range in degrees:  –30° to +90°**

# EARLY, LATE OR ON TIME?
## TRANSITION ZONE

### What Is the Transition Zone?

It is simply the orientation of the heart's electrical axis in the horizontal plane. You could say the twin of the QRS axis. The difference is that we use the precordial leads to determine it, viewing the body from above.

The transition zone occurs in the lead where the QRS complex shifts from negative to positive. In the illustration, you can see two possible scenarios.

**Scenario A:** The shift from negative to positive occurs exactly at V3. In this lead, the positive and negative waves are about the same size. This is referred to as transition at V3.

**Scenario B:** In V3, the QRS complex is still predominantly negative, while in V4 it is already mostly positive. The transition therefore takes place somewhere in between. In our report, we would write: transition at V3/V4.

**The normal transition zone lies at V3, V4, or between the two.** Anything outside this range is considered either early (V1 and V2) or late (V5 and V6).

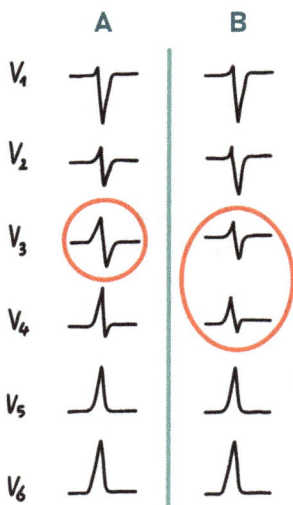

Two scenarios: At first, the QRS complex is predominantly negative. But at a certain point, it changes its orientation to positive.

> **Don't Get Confused:** Sometimes in clinical practice the term rotation is used instead of transition. However, the two are not synonymous. The transition defines the degree of rotation. But we only speak of a rotation when the transition deviates from its normal value.
>
> Since we like to keep things simple, just forget about the term rotation for now. The transition is all you need.

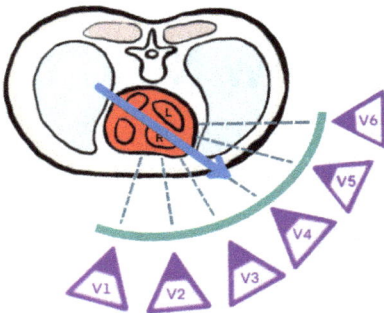

With transition at V4, the direction of impulse spread in this heart would look roughly like this. It broadly matches the orientation of the septum.

**Tip!**
For our report, it's perfectly sufficient to simply note transition + lead.

As with the QRS axis, the transition is difficult to interpret, since it also provides only very vague clues. As long as you know how to determine it and where the normal range lies, you've done your part.

The turning point is located in the lead toward which most of the electrical charge is mainly directed. The transition zone - just like the QRS axis - depends on the anatomical position of the heart, its dimensions, and the condition of the conduction system.

# EVERYTHING AT A GLANCE

Look at the precordial leads: The QRS complex is usually negative in V1 and positive in V6. Find the lead where the QRS complex "flips" from negative to positive.

This can happen either exactly in a single lead or between two neighboring leads. That's where the transition zone is.

**Normal range:** V3, V4, or between the two.

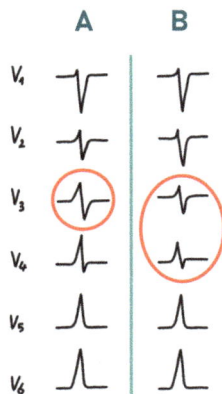

# THE REPORT - AND HOW NOT TO SCREW IT UP

So, things are starting to get real. Now that you've taken a close look at the cardiac cycle and both electrical axes, it is time to hand you the tools you'll need to systematically interpret an ECG.

One of the biggest problems for many students is that, while they do understand the basic principles of the ECG, they don't know where to actually start with the interpretation or what exactly to focus on. That's something we definitely want to avoid!

By the end of this chapter, you'll know what a good ECG report has to include and how to create one step by step.

## Keep the Big Picture in Mind

> **Lesson 1:** "Skip a lead, and your patient's heart might stop to beat."

It's absolutely essential to include all leads in your interpretation. Looking only at the limb leads and calling the heart healthy is a serious mistake.

That would be like trying to sell someone a car based only on a photo from the front. You wouldn't notice that the back end is completely wrecked unless you also took a picture from the side. Understandably, anyone would be pretty upset if you handed them this so-called "perfect report", only to later find out they need to buy a new trunk.

The same principle applies to the ECG. The example on the next page shows that you could even miss a heart attack if you make this mistake.

The precordial leads let us peek around the corner and spot the damage at the back.

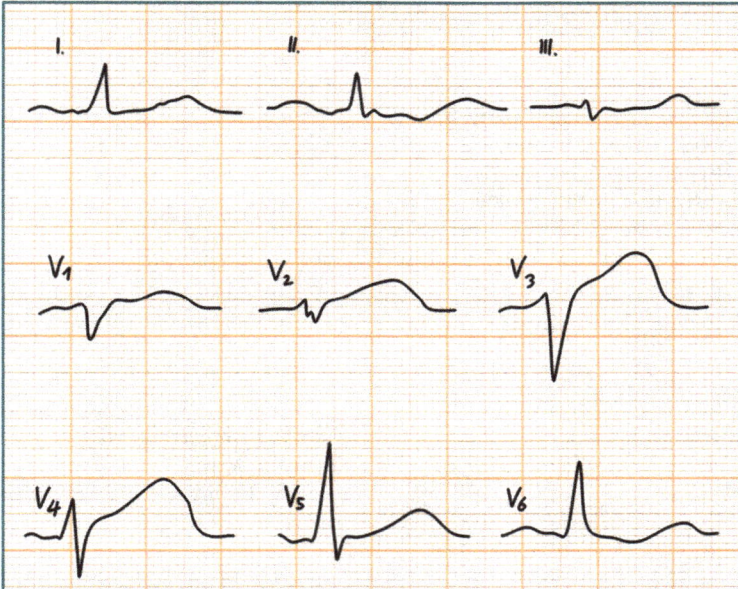

Leads I, II and III look almost unremarkable here. The same goes for aVR, aVL, and aVF (not shown due to space). The fact that this person is in real danger only becomes clear when you look at leads V1–V4!

In these leads, the segment between the QRS complex and the T wave is elevated. It is sitting above the isoelectric line we have to imagine. This is called ST elevation, and it's a classic sign of a myocardial infarction. We'll cover that in a later chapter.

## Where Do I Start With the Interpretation?

Now that you know you always have to keep an eye on the big picture, we can move on to actually analyzing the ECG.

If we want to find out whether everything is fine, there are two things we have to pay close attention to: unusual changes in shape and the timing of certain parts of the ECG.

To make sure nothing important gets overlooked, we follow a specific approach every time. That way, you can be confident you won't miss anything essential.

And so you don't have to memorize the order in a dry, boring way, you can remember the process through a little story.

## A Road Trip Through the Mountains

**1.** Careful preparation is key before setting out on a long trip. First, we want to make sure our car is actually roadworthy and up for the journey. So we pop the hood and take a quick look: How high is the RPM, and is the engine running smoothly without stuttering?

**Heart rate and rhythm (bradycardic/ tachycardic? regular/irregular?)**

HR? Rhyth.?

Height

P Wave Duration

**2.** Once we've got the overview, we hit the road. According to our map, the first stop is the P hill, which we have to cross before we can roll down into the next valley.

When we see it rising up ahead of us, we're relieved, because that means we're still on the right track. At the top, we pull over for a quick bathroom break.

Luckily, the P hill isn't too steep or bumpy, otherwise parking would be a real hassle. While we're there, we take a moment to admire its size and esti-mate how long it takes to get across.

- P wave present regularly?
- Shape? (peaked? notched?)
- Amplitude? (= Height)
- P wave duration <100 ms?

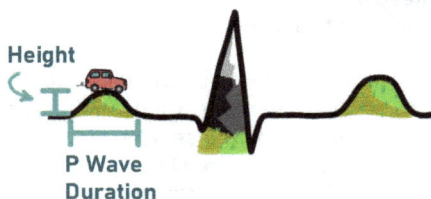

**3.** On we go, down into the valley and straight toward the massive QRS mountain range. At the foot of the mountain we pull over for another bathroom break - our bladder isn't exactly the strongest.

We glance back and check the clock: how long did it take us to get from the start of the P hill to the beginning of the QRS mountains?

**PR interval between 120 and 200 ms?**

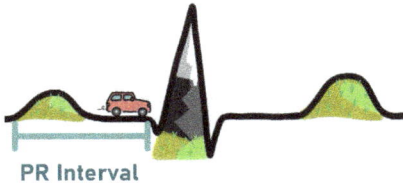

**PR Interval**

**4.** Now we're racing up the steep mountain road, and we reach the summit faster than we expected.

We just can't resist pulling over to en-joy the view. This time we take it slow and examine everything in detail.

What's the orientation of the mountain when you look at it head-on? How is it aligned from a bird's-eye view? How tall is it? What shape does it have? And of course, we also think about how long it takes to cross it.

- **QRS axis?**
- **Transition zone?**
- **Amplitude?**
- **Any abnormal changes in shape?**
- **QRS duration <110 ms?**

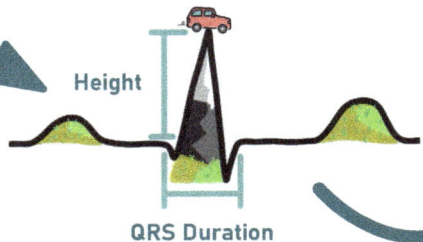

**Height**

**QRS Duration**

**5.** Back down in the valley, we can't help but gaze out the window and admire the landscape.

We're relieved that the valley ahead is completely flat, because if it weren't, our engine wouldn't get a chance to recover from the climb and might overheat.

**ST segment flat? (on the isoelectric line, no elevations or depressions)**

**6.** Finally, we still have the T hill to cross. It's a bit bigger than the P hill, so we pay extra attention to its shape.

Once we reach the end, we wonder how long it would take to drive back to the start of the QRS mountains, because that's where our lodge is, and we want to make it back before nightfall.

- **Shape of the T wave (normal? flattened? inverted?)**
- **Amplitude?**
- **QT interval?**
- **Should be taller than the P wave**

If you need stories like this as a memory aid, it's probably only in the beginning. With a bit of practice, you'll soon know by heart what to include in your report.

**Tip!**
Just start all the way on the left, at the beginning of a cardiac cycle, and work your way to the right. At each section, ask yourself what you need to pay attention to. **The cardiac cycle itself serves as your visual guide for the report, making sure you don't miss anything.**

And there's no shame in carrying a little cheat sheet with the interpretation approach in your pocket at the beginning.

## The 1–2–1 Rule for Normal Values

In everyday clinical practice, there's no getting around memorizing a few values. Fortunately, it's pretty straightforward if you just keep these three numbers in mind:

1-2-1

Behind these three subtle numbers lies a clever system that isn't so obvious at first glance.

It helps you keep track of the upper limits of the three most important values: **the P wave duration, the PR interval, and the QRS duration.**

**Here's how it works:**

- What's obvious: Each of the three numbers represents the first digit of the upper limit of the value.

| 1 | 2 | 1 |
|---|---|---|
| **100** | **200** | **110 ms** |
| **P** | **PR** | **QRS** |

- The QRS duration is a bit of an outlier, with 110 instead of 100 ms. Just remember that this corresponds to the second "1" in the memory aid. So it makes sense that the associated value also contains the digit 1 twice.

- The values follow the same order in which they appear during the cardiac cycle. Alternatively, you can remember the sequence by counting the number of letters (1-2-3).

- The middle value, the PR interval, is the only one that also has a relevant lower limit. To remember it, just drop the dashes in your mind and pull the numbers toward the middle.

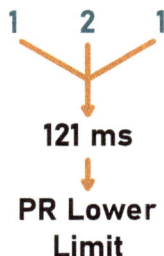

1  2  1

**121 ms**

**PR Lower Limit**

Strictly speaking, it's 120 ms, but we can get away with that one millisecond so the memory aid still works.

## What About the QT Interval?

Diese ist stark frequenzabhängig. DThis one is highly rate-dependent. That's why there are tables that let you look up the expected QT interval for a given heart rate.

These days, however, it's more common to take the measured QT interval and plug it into a formula together with the heart rate, which "corrects" for the rate dependency. The result is a value that always has the same reference range. **The so-called corrected QT interval (QTc).**

Most ECG software and even many paper ECG machines will calculate the QTc for you automatically. It should not exceed the following value:

$$QTc \leq 450 \text{ ms}$$

**Here we bend the rules a little:** women are actually allowed a value of up to 460 ms. But it's fine to just remember the cutoff of 450 ms.

That way, we're being a little more cautious with women, but we avoid the risk of mixing things up.

**Tip!**
In fact, a shortened QT interval is clinically of little relevance. What really matters is keeping an eye out for prolongation.

While 450 ms for men and 460 ms for women serve as general thresholds for a prolonged QTc, **clinical concern typically peaks when QTc exceeds 500 ms.** At this level, the risk for life-threatening arrhythmias increases substantially.

**A Quick Look Is Enough!** With the following method, you can get a rough estimate in no time.

Simply look at the distance between two neighboring R waves. Split it right down the middle and imagine a vertical line there. The T wave should lie in the left half and not touch that line.

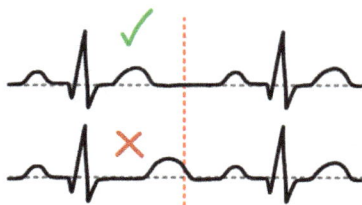

Halving Rule: The T wave should lie entirely in the left half.

If the line falls within the T wave, you can assume the QT interval is too long. The same applies if the T wave crosses the line completely and ends up in the right half.

**Important!**
This rule of thumb no longer applies at a heart rate above 100 per minute!

## How Detailed Does a Report Need to Be?

Voltage measurements (the amplitude of P, QRS, and T) are usually only important in more specific situations, and you don't necessarily have to include them in your report.

They can often serve as a clue to hypertrophy, but the ECG diagnosis of hypertrophy or dilatation has been losing importance in countries with good medical resources, simply because echocardiography is so widely available.

The rest of the values, however, must never be left out!

> **A normal report would sound something like this:**
>
> "Normofrequent sinus rhythm, P wave duration and morphology unremarkable, PR interval within normal range, QRS axis normal, transition zone in V3/V4, QRS duration and morphology unremarkable in all leads, no ST segment elevations or depressions, QT interval normal, T waves unremarkable."

You'll notice that you can summarize things quite a bit. If all the individual points were normal, you can simply note that briefly and to the point.

Most of the time, only the values that were abnormal are mentioned explicitly. Example: "QRS in V1/2 prolonged to 120 ms."

**Tip!**
The term **sinus rhythm** actually means two things:

First, that P waves are present with every cardiac cycle. If they weren't, you couldn't call it a sinus rhythm, since P waves originate from the sinus node.

Second, it means that everything is running nice and rhythmically. So with this single term, you've killed two birds with one stone.

## How Do I Get the Necessary Measurements?

Most of the time, ECGs are screen-based. That means you can display the leads flexibly, adjust the speed, and much more. The values are usually calculated and displayed for you right away.

These days, most paper ECG machines also measure automatically and print the results.

In both cases, you should briefly double-check the automatically generated values, since the device can, of course, make mistakes. On a computer, you can do this easily with the built-in tools.

But if you're dealing with a paper ECG, you have two methods at your disposal.

- **Method 1 – The ECG Ruler:**

Back in the day, it was a must-have in every white coat pocket. Today, in the era of screen-based ECGs, it's almost a rare collector's item. If you ever need one, you'll usually have to sneak it from an old-school colleague.

Using it feels more complicated than actually finding one - but only at first glance. Once you've taken a closer look, it's really quite simple.

The best way is to have someone show you how to use it, since a written manual wouldn't be all that helpful. You can also find plenty of demonstration videos online.

**Important!**
A ruler gives you different scales to read off everything you need. The only thing you have to watch out for is the speed at which your paper ECG was recorded. It's either 25 mm/s or 50 mm/s. This information is always noted somewhere at the margin of the ECG.

When you hold the ruler up to the ECG, make sure you use the scale that matches the speed of your tracing.

- **Method 2 – Counting Boxes:**

This is the stripped-down survival method. It is less accurate, requires a bit of mental math, and is error-prone if you're not paying close attention.

**First, check the speed of your ECG. Often it's 25 mm/s.** That means one second of ECG is spread across 25 mm of paper. **From that, you can calculate that 1 mm on paper equals 40 ms** (20 ms at 50 mm/s).

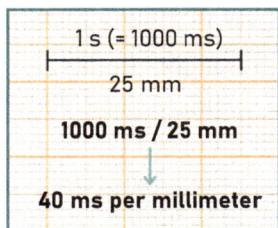

```
1 s (= 1000 ms)
├─────────────────┤
     25 mm

 1000 ms / 25 mm
        ↓
40 ms per millimeter
```

Conveniently, ECG paper always comes with a 1-mm grid. But with the naked eye, your accuracy is limited to about half a box, which naturally leads to some inaccuracy.

When it comes to amplitude, it's pretty much the same. Every ECG starts with a so-called calibration spike, a rectangular wave that represents exactly 1 millivolt (mV, voltage). It serves as a reference to figure out how many millimeters correspond to 1 mV.

Then, simply divide 1 mV by the number of boxes. Or, if you look around, you'll often find it written out as well. Most commonly, it's 10 mm/mV.

---

**Here, the calculation would be:**

**1 mV / 10 mm = 0.1 mV per mm**

---

Once you know both values, you're ready to get started. On the next page, you'll find an example. Pay attention to just how imprecise this method can be in practice.

If you count the boxes of the P wave in the example below, you get an amplitude of 0.2 mV (2 boxes) and a duration of about 160 ms (4 boxes).

As you may notice, the duration is up for debate. Some might be a bit less generous here and count 3.5 boxes instead. That would give 140 ms - a pretty big difference. But in this case, either way it would be pathological.

In borderline cases, however, this inaccuracy can easily lead to one person classifying a finding as pathological while someone else still considers it unremarkable.

So much for interpretation. Hopefully, this slightly longer chapter has shed some light on the subject and you now feel less lost when you're asked to look at an ECG and interpret it.

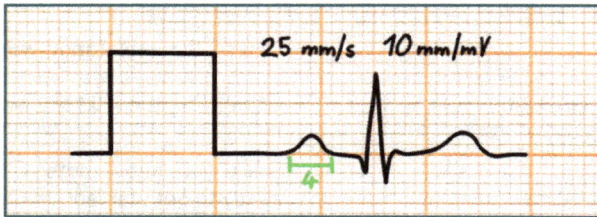

Calibration spike with one cardiac cycle. The ECG is slightly enlarged for better clarity.

**Tip!**
Finding pathology on the ECG is detective work. We gather clues and interpret them, ideally alongside the patient's history and other examination findings.

For example, the QRS duration is prolonged in several conditions. So don't jump to the first diagnosis that comes to mind; take your time and weigh what argues for and against each possible pathology.

**In the end, you draw your conclusion.**

**Note:** With the exception of the practice examples in the last chapter, all ECGs in this book use 25 mm/s and 10 mm/mV! This way you won't run into unnecessary confusion at the beginning.

# EVERYTHING AT A GLANCE

**Your Report Should Include:**

1. Heart rate
2. Rhythm: regular/irregular?

**From here on, follow the cardiac cycle from left to right!**
3. P wave
   - Present regularly?
   - P wave duration
   - Morphology (peaked? notched?)
   - Amplitude
4. PR interval
5. QRS complex
   - QRS axis
   - Transition zone
   - QRS duration
   - Morphology
   - Amplitude
6. ST segment (should be isoelectric)
7. QT interval
8. T wave
   - Morphology (inverted? other abnormalities?)
   - Amplitude (should be higher than the P wave)

**Reference Values – Remember the 1–2–1 Rule!**
- P wave duration <100 ms
- PR interval 120–200 ms
- QRS duration <110 ms
- QTc interval ≤450 ms (Halving Rule)

**Counting Boxes:**
- 25 mm/s = 40 ms per box
- 50 mm/s = 20 ms per box

# BASIC UNDERSTANDING
## DILATATION AND HYPERTROPHY

From this point on, we'll start working with our own system to help you understand things more clearly. To give you a better feel for the dynamics of the ECG, we'll go over some basic mechanisms using dilatation and hypertrophy as examples.

Both are pathological morphological changes (meaning changes in the structure of the heart muscle) that occur mainly in older patients and are things you'll frequently come across in everyday clinical practice.

**Dilatation and hypertrophy - you've probably heard both terms before. But what exactly is the difference between the two?**

## Dilatation

This refers to a remodeling process of the heart in which the muscle becomes thinner, develops scarring, and loses strength. The affected part of the heart essentially becomes "floppy."

**You can picture it like this:** When you take a brand-new balloon out of the package, it's tight and compact. Blow it up and then let the air out, and it quickly shrinks back down, with its tension pushing the air out through the opening.

But if you repeat this several times - or leave the balloon inflated for a few days before letting the air out - it looks different afterward. Its wall is thinned and stretched, and the balloon as a whole is larger, even though there's no pressure inside anymore. It no longer has the tension it needs to collapse completely and push the remaining air outward.

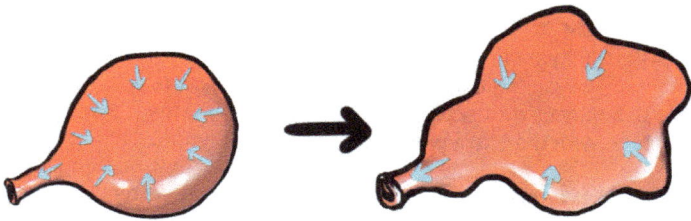

The balloon on the left represents a healthy heart wall. On the right, you can see how the wall has been stretched out by dilatation, making the balloon larger.

Morphologically, a pathologically dilated heart behaves like the balloon on the right. Besides chronic volume overload, the most common causes are actually infectious myocarditis (inflammation of the heart muscle), coronary artery disease with ischemia (reduced blood supply), or a genetic predisposition. These are what trigger the harmful remodeling processes.

Once the heart muscle has become overstretched, the same happens to the tissue of the conduction system. As a result, the electrical charge takes longer to pass through the muscle - simply because the path has become longer. On the ECG, the wave is wider than usual, since the extra time is reflected on the x-axis.

In this example, the left ventricle (from the patient's perspective) is dilated. **Its conduction pathway is therefore considerably lengthened compared to the other ventricle, and lead A records the "noise" of the charge rushing through for much longer.**

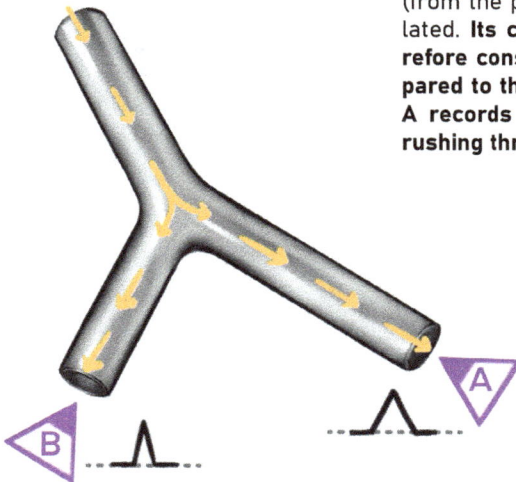

With dilatation, we can observe the elongation of the muscle on the ECG.

## Hypertrophy

In this case, the heart is stimulated by chronic, non-physiological pressure overload to build more muscle tissue in order to cope with the extra workload.

Such persistent pressure overload occurs most commonly in people with valve stenosis (narrowing of a heart valve), since a smaller valve opening means more work for the heart. Chronic high blood pressure is also one of the most frequent causes.

**You can picture the effects on the ECG like this:** As the mass of the heart muscle increases, the conduction pathways also thicken, allowing a larger electrical charge to pass through. This charge can then generate the additional force needed to move the thickened muscle.

Another word for the force of a charge is voltage - and voltage influences the y-axis of the ECG. Increased voltage can therefore be a clue to hypertrophy. On the next page you'll find an example.

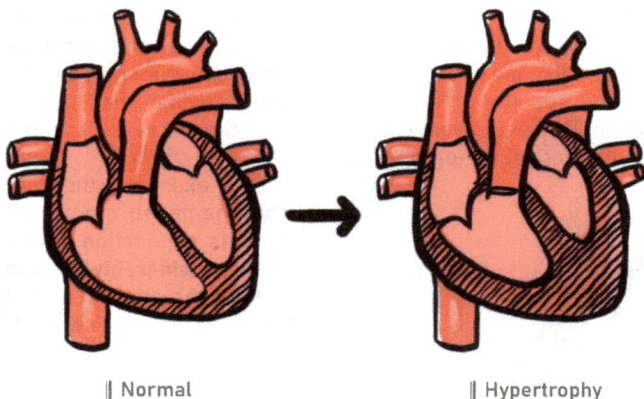

‖ Normal                    ‖ Hypertrophy

**Important!**
Although we can make certain statements about the morphology of the heart, an ECG can never tell us how "weak" or "strong" the heart is actually pumping. For questions like that, imaging methods such as echocardiography are required.

In this example, only the left ventricle is hypertrophic. A larger amount of electrical charge flows through the thickened conduction pathway, carrying more force and therefore creating a much louder "sound." Compared to the right ventricle, we see a clearly higher deflection on the ECG.

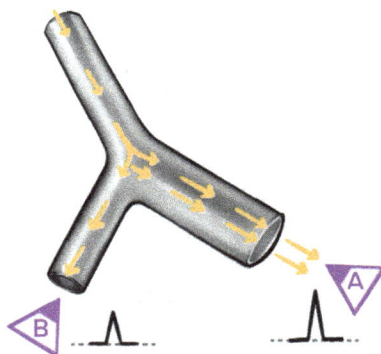

On the ECG, hypertrophy shows up as an unusually high deflection due to the increased muscle mass.

## Nice to Know – Healthy vs. Unhealthy

In well-trained endurance athletes, a so-called eccentric hypertrophy develops over time. The heart muscle initially dilates to allow more blood volume into the chamber, while at the same time the muscle wall thickens proportionally. This prevents negative effects. In other words, by combining dilatation with hypertrophy, a physiological dilatation is achieved that preserves the pumping power of the heart. As a result, the same amount of blood per minute can be circulated with fewer beats. During a marathon, for example, the heart rate is therefore lower.

In contrast, with non-physiological stress - such as a backlog caused by valve stenosis - the heart muscle usually grows very quickly. The growth of the blood supply cannot keep up with this speed. A larger muscle requires increased blood supply - and this demand rises even further during physical exertion. At a certain point, the muscle becomes so large that it can no longer be adequately supplied by the blood vessels, leading to ischemia (lack of oxygen). This can produce symptoms very similar to a heart attack and, in the worst case, just as severe consequences.

## The Big Picture

In this chapter, all examples used the fictional leads "A" and "B," which were located right next to the conduction pathways for easier understanding. This way, they "listened" to those signals in isolation and each recorded only a single wave.

The real leads, of course, are located on the outside of the body. Because they record from a greater distance, their "field of view" is broader. **They capture the heart's entire conduction process, recording the complete cardiac cycles.**

Depending on where a lead is located on the body, the "perspective" on the heart and its conduction system changes.

The two examples below are meant to show you what happens when leads are located at different sites on the body.

Pay close attention to how the QRS complex changes depending on the lead's location.

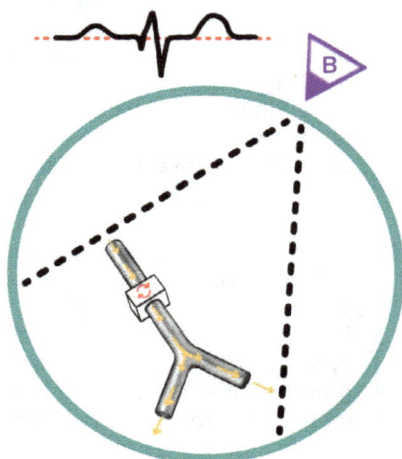

Lead A is oriented exactly against the flow of the entire system, so most of the charge rushes toward it. The QRS complex is predominantly positive.

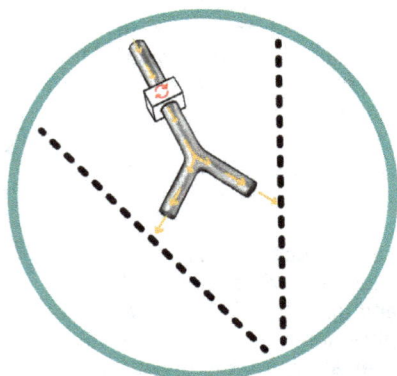

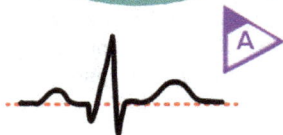

Lead B, on the other hand, is located laterally. Most of the charge moves perpendicular to it, so the QRS complex appears more balanced.

## Ventricular Chaos

In the image below, you see the actual spread of excitation in the heart shown in yellow. Above it, in blue, is the simplified version we imagine in order to match our power plant system.

When you compare the actual spread of excitation in the ventricles with that in the atria, you'll notice that in the ventricles the excitation changes direction drastically. It moves across the septum toward the apex of the heart, then back upward to spread over the walls. Along the way, small portions of the excitation branch off continuously.

No matter how a lead "looks" at the heart, some parts of the ventricular excitation are always moving toward it, while others move away. As a result, the ECG section that reflects ventricular excitation – the QRS complex – always contains both positive and negative components.

Which parts are perceived as positive and which as negative depends heavily on the "angle of view" of the lead on the heart. That's why the QRS complex looks different in every lead.

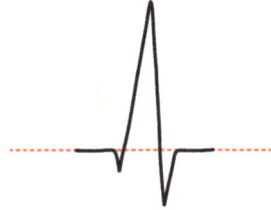

The QRS complex can be predominantly positive, negative, or balanced, depending on the location of the lead.

**Tip!**
The spread of excitation shown in blue may seem overly simplified, but for most of the pathologies in this book, thinking of it this way is more than enough. In fact, this perspective can even help you understand certain problems more easily.

Now that we've finally covered the basics, it's time to move on to the really interesting part. We'll start all the way on the left with the P wave and, chapter by chapter, stop at every pathology we encounter along our journey through the entire heart mountain range.

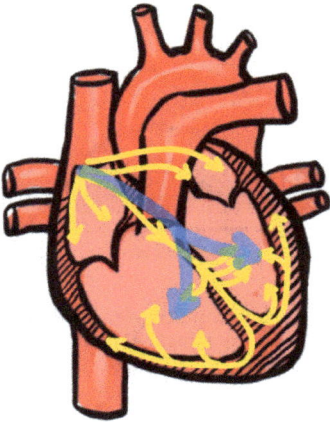

The actual excitation spread is shown in yellow, and the simplified version in blue.

# MAJESTIC MOUNTAIN SCENERY
## ATRIAL MORPHOLOGY

We'll start with the P wave. Its morphology mainly gives us clues about structural changes in the atria. This is fairly nonspecific and serves only as a hint toward possible dilatation or hypertrophy. A definitive diagnosis, however, can only be made with imaging.

Because echocardiography is now so widely available, detecting atrial pathologies via ECG has lost much of its clinical importance. Still, it's worth running through briefly, since it helps with your overall understanding.

As you already know, excitation takes longer to travel through a dilated muscle than through a healthy one, because the overstretched muscle represents a longer path that needs to be covered.

A hypertrophic atrium, on the other hand, produces a "louder signal," since its muscle is thicker and can conduct more charge at once. The duration, however, remains normal.

> **Morphological changes in the atria are typically best observed in lead II.**

**From this illustration, you can see how it shows up on the ECG in practice:**

Representation of the two atria as mountain peaks. The effects of pathological atrial changes on the P wave are explained on the next page.

## 1. Physiological Atria

Since the diagram is shown in a magnified view, the first image serves as a reference for a normal P wave so you can compare the others to it.

Beneath the P wave you'll see our two atrial "mountains." They represent the left and right atrium. Viewed from the front (ventrally), the right atrium lies in front of the left and partly overlaps it.

Normally, both are about the same size, and you could easily draw a bridge between their peaks. That's what gives the P wave its typical shape.

## 2. Right Atrial Hypertrophy

The right "mountain" grows larger and gradually covers more of the left, until the left one is almost completely hidden. It no longer contributes visibly to the overall shape. As a result, the curve follows only the right peak, making it taller and sometimes a bit sharper.

## 3. Left Atrial Hypertrophy

The left "mountain" grows larger and gradually pushes out from behind the right atrial peak, eventually dominating the silhouette of the range.

The right peak, however, remains visible in front. This creates the classic double-humped P wave. Usually, the second hump is taller than the first, though in some cases they can be about the same height.

## 4. Dilatation

No analogy is needed here to remember it. Any P wave lasting longer than 100 ms points to a dilatation. You've already got that cutoff memorized thanks to the 1-2-1 rule.

At this point, though, you can't distinguish between the right and left atrium.

If you also spot one of the suspicious P-wave shapes and, on top of that, see a prolonged P-wave duration, that suggests a combination of hypertrophy and dilatation.

**Tip!**
You shouldn't jump straight into your report saying you've found an atrial pathology. Instead, just describe the P-wave changes you've observed.

That kind of description is much more useful, since you can't make a definitive statement about cardiac morphology from the ECG alone anyway. What it can do, however, is prompt further work-up, with imaging if needed, to get to the root cause.

And with that, you've done your job.

### Nice to Know – Atrial Conduction

The reason this mountain analogy works is rooted in anatomy. Since the sinus node sits in the right atrium, depolarization starts there before spreading to the left.

And because events that happen earlier show up more to the left on the ECG, the first half of the P wave actually reflects right atrial activation, while the second half corresponds to conduction through the left atrium.

## Here's what it looks like in real ECGs:

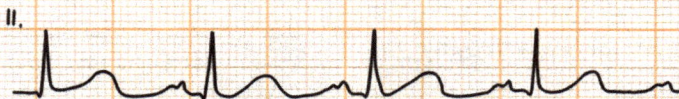

Here's a classic example of suspected left atrial overload. You can see the typical double-peaked shape and a P-wave duration of about 100 ms. This means left atrial hypertrophy without dilatation. This pattern is also known as **P mitrale**, since left atrial overload - and therefore hypertrophy - is often caused by a narrowed mitral valve. **A simple way to remember it: think of the double-peaked P wave as an "M."**

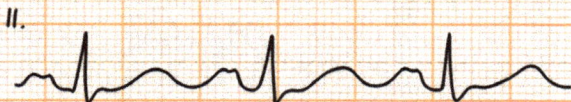

In this example, the P wave also shows signs of dilatation. While there's still a double-peaked shape, the more striking finding is the clearly prolonged P-wave duration of 160 ms.

II.

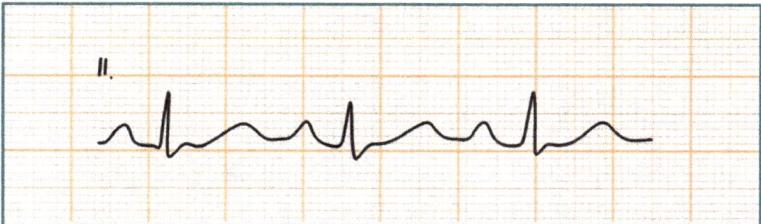

Finally, here's a nice example of a pointed P wave. It shows clearly what this can look like in practice. In this case, the right atrium is likely enlarged. The P-wave duration is also outside the normal range. This finding is called **P pulmonale**, since it's typically caused by **increased resistance in the pulmonary circulation (pulmo = lung).**

# EVERYTHING AT A GLANCE

1. **Atria of normal size**

2. **Peaked, tall P waves** = right atrial hypertrophy (P pulmonale)

3. **Double-peaked P waves** = left atrial hypertrophy (P mitrale)

4. **P wave longer than 100 ms** = dilatation

Combinations are also possible.

Changes in P-wave morphology are only clues, definitive diagnosis requires imaging!

**These changes are best seen in lead II.**

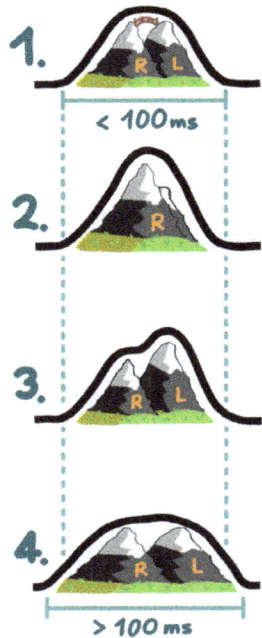

# CHAOS IN THE ATRIA
## ATRIAL FLUTTER AND FIBRILLATION

For all sorts of reasons, the sinus node's signals can get drowned out by rapid, chaotic impulses, causing the atria to completely lose their rhythm.

These impulses don't come from the sinus node itself but from outside of it. We distinguish between two types of this loss of control, each with a different origin.

### Atrial Flutter

If the atrial rate is simply too fast but there is still some degree of regularity in the signals, we call it atrial flutter.

In many cases, these faulty impulses are caused by what's known as a re-entry mechanism. That means the ventricular depolarization loops back up into the atria. That's something the "AV charging station" is actually supposed to prevent.

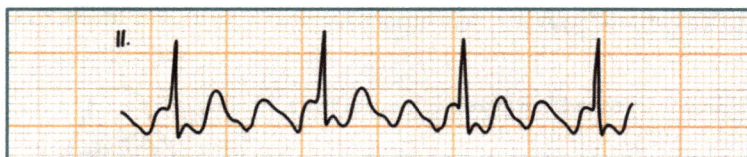

These sawtooth-like waves between the ventricular complexes are the atrial impulses. There can be two or more of these waves between each QRS complex. In this case, you can even see a third wave starting up each time, only to be cut off by the ventricular complex.

### Nice to Know – Filter Function

The "AV charging station" needs a brief pause each time before it can pass on the next impulse. And that's a good thing, because otherwise every single flutter impulse would trigger ventricular activation. In case of the example shown above, the atrial rate would be around 300, far too fast for the heart to refill properly after every beat.

## Atrial Fibrillation

In the second scenario, the impulses lose any trace of regularity, leaving nothing but pure chaos. That's what we call atrial fibrillation.

The cause is chaotic trigger impulses that originate near the openings of the pulmonary veins in the left atrium and spread out from there.

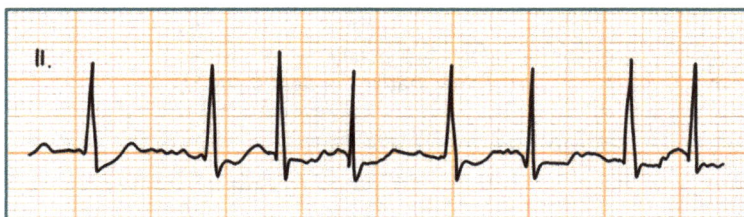

You can recognize atrial fibrillation by two clear signs. First, there's no regularity left in the PR segment, distinct P waves are no longer visible, and all you see is pure chaos. Second, the spacing between QRS complexes is irregular, because atrial fibrillation is always accompanied by an arrhythmia.

### Nice to Know – Why Arrhythmia?

The ventricular arrhythmia in atrial fibrillation happens because the atrial impulses are so small and irregular that only some of them are strong enough to trigger ventricular activation. How often that occurs is basically random. The "AV charging station" filters out the weaker impulses and blocks them. So once atrial activation loses its regularity, the ventricles inevitably become irregular as well.

**But keep in mind:** arrhythmia doesn't necessarily mean bradycardia. Even if only a fraction of the many fibrillatory impulses gets through, an irregular ventricular rate of over 100 beats per minute is still very possible.

### Important!

If the ECG signal quality is poor, interference can make the tracing look shaky or irregular. Sometimes it might even resemble a fibrillation pattern. However, in that case, the QRS complexes would still appear rhythmically, since it's merely an artifact, not an actual atrial fibrillation.

**Important!**
As you can see from these examples, the filter mechanisms of the "AV charging station" play a crucial role.

On the one hand, they limit how often conduction can occur, preventing the heart rate from climbing too high. On the other hand, impulses that are too weak don't make it through to the ventricles.

If these protective functions are bypassed, it can lead to serious problems. You'll learn more about that in the next chapter.

## Nice to Know – What to Do About Atrial Chaos?

Atrial flutter often progresses to atrial fibrillation if left untreated. Both rhythm disturbances can be managed with medications (antiarrhythmics), electrical cardioversion, or catheter ablation.

Since the atria are no longer pumping effectively, blood clots (thrombi) can form. If one breaks loose, it may block a vital blood vessel (embolism). That's why anticoagulation (commonly called "blood thinning") is usually required.

In addition, prolonged fibrillation can weaken the heart and lead to heart failure. The treatment is the same as for flutter, but atrial fibrillation tends to recur or even become persistent in many patients.

# EVERYTHING AT A GLANCE

- **Atrial Flutter**
Sawtooth pattern of P waves with a recognizable regularity.

- **Atrial Fibrillation**
Chaotic atrial activity without regularity, plus ventricular arrhythmia. No identifiable P waves.

# AV STATION SHORT CIRCUIT
## WPW SYNDROME

We're leaving the P wave behind and moving on to the PR segment and PR interval. One interesting pathology in this area is Wolff-Parkinson-White (WPW) syndrome. It's a bit of a niche topic compared to the other conditions in this book, but it's incredibly helpful for understanding the AV station.

WPW syndrome develops through a short circuit in the AV charging station. Think of it as inexperienced maintenance staff accidentally connecting two wires that were never meant to go together. The diagram below shows you how to picture it.

And of course, that mistake comes with a few consequences. Part of the impulse rushes through the extra wire, bypassing the AV charging station. And that's exactly what the leads can "pick up."

Normally, the station holds the impulse for a brief moment before letting it through. That's why there should be a "silent" gap between the P wave and the QRS complex. But with this extra pathway, part of the impulse is not held back and a signal appears earlier than it should, shortening the PR interval.

We call this the delta wave, because the area under it looks a bit triangular - and with a little imagination - like the Greek capital letter Delta (Δ).

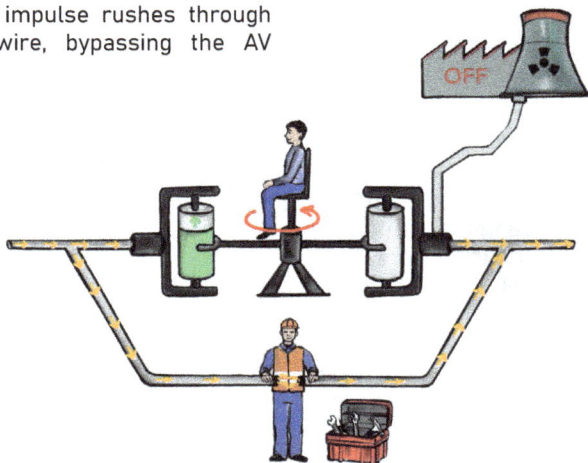

One careless move, and there it is: the impulse splits, and part of it rushes past the station unchecked.

Delta Wave

<120 ms

⚡ + Fantasy = Capital Delta (Δ)

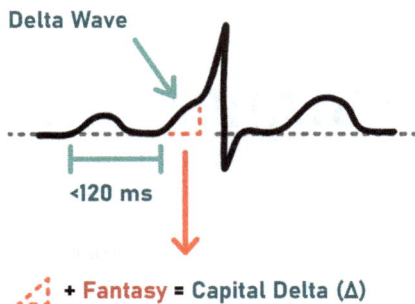

This image shows the textbook version of a delta wave. But it doesn't always look like this.

Sometimes it's narrower, and in some cases it may even stand alone instead of merging directly with the QRS complex, as shown above.

Occasionally, it can also be very tall and narrow, hugging the QRS complex so closely that the whole thing looks like one large, broad R wave. Delta waves that hide like this are often overlooked or misinterpreted by beginners.

## Significance of the Delta Wave

Because part of the impulse bypasses the AV charging station, the delta wave represents an early ventricular depolarization. In other words, you could say it actually belongs to the QRS complex.

Fortunately, we don't have to rely on spotting the delta wave alone to make the diagnosis!

Since the PR interval is measured from the beginning of the P wave to the beginning of the QRS complex, it's logically shorter here, because ventricular activation starts earlier.

**That's our main clue - the PR interval drops below the lower limit of 120 ms.**

Only when you notice that during interpretation should you go hunting for delta waves! If the PR interval is over 120 ms, so within the normal range, a WPW syndrome is unlikely.

On top of that, the QRS complex is usually wider, since the delta wave "adds itself in." That's why **the QRS duration often exceeds 110 ms.**

**Important!**
A delta wave doesn't have to show up with every beat. It may only appear occasionally in some beats on the ECG. The other beats will then show a normal PR interval again.

This is called an intermittent WPW syndrome!

## Nice to Know – WPW in Clinical Practice

The short-circuit cable is formally called the Kent bundle or **accessory pathway (AP)**. In about 50% of cases, such a pathway is symptom-free and requires no treatment.

**It's only when symptoms occur that we speak of WPW syndrome.** Typical symptoms include paroxysmal tachycardia, dizziness, syncope, chest pain, and palpitations.

Tachycardia episodes can be interrupted and prevented with medications, but the causal treatment of WPW syndrome is catheter ablation of the accessory pathway, which is cut through in a minimally invasive procedure.

### Important!

Treatment of a symptomatic accessory pathway is mainly recommended because of the potential problems it can cause down the line.

Since the AP bypasses the AV node, its filtering function is lost. If atrial fibrillation develops later in life - as it does in many people - it can spread unchecked to the ventricles and trigger life-threatening ventricular fibrillation. In healthy individuals, without such an extra pathway, that wouldn't happen.

It's also possible for part of the ventricular activation to travel back through the AP into the atria and from there re-enter the ventricles, creating a loop that can lead to a severe tachycardia (reentry tachycardia).

## Here's what it looks like in real life:

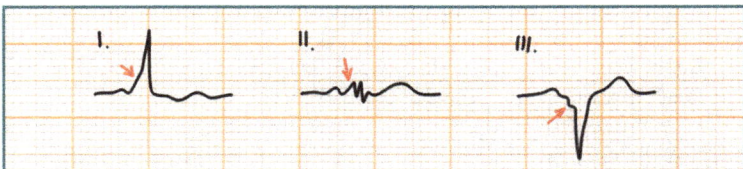

Let's start with a fairly classic example. The red arrows point to the early ventricular activations (delta wave, pre-excitation), which show up as little bulges on the side of the R wave. In lead II, you can also see what a "standalone" delta wave looks like when it doesn't merge with the QRS complex. The PR interval is clearly below 120 ms, and the QRS width exceeds the cutoff of 110 ms.

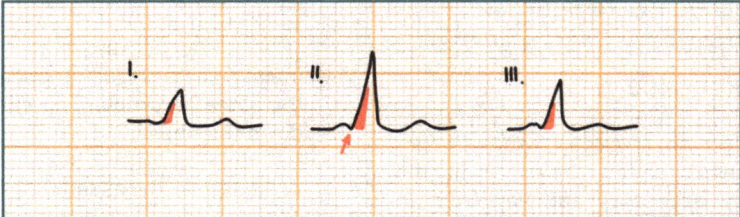

Here's a tricky case. As mentioned earlier, delta waves can also be tall and pointed, even blending seamlessly into the QRS complex. The typical "delta bulge" is missing here. The section corresponding to the delta wave is marked in red. Its giveaway is that it starts immediately after the P wave - skipping the usual "quiet gap" (see arrow) - which makes the PR interval extremely short.

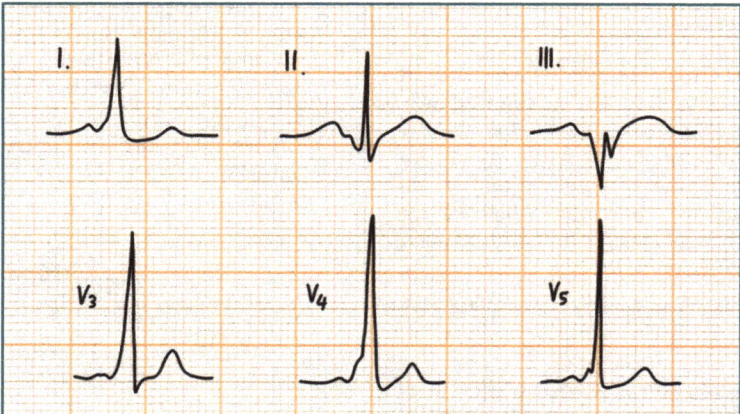

Finally, here's an example without any visual aid.

Apart from lead I, the first three leads don't really show anything suspicious that would suggest WPW syndrome. But in V3 through V5 you can clearly see the delta wave becoming more pronounced, until in V5 it even forms a small notch.

Again, the PR interval is too short in all leads. You can spot this without measuring, because the QRS complex starts right away after the P wave.

AP

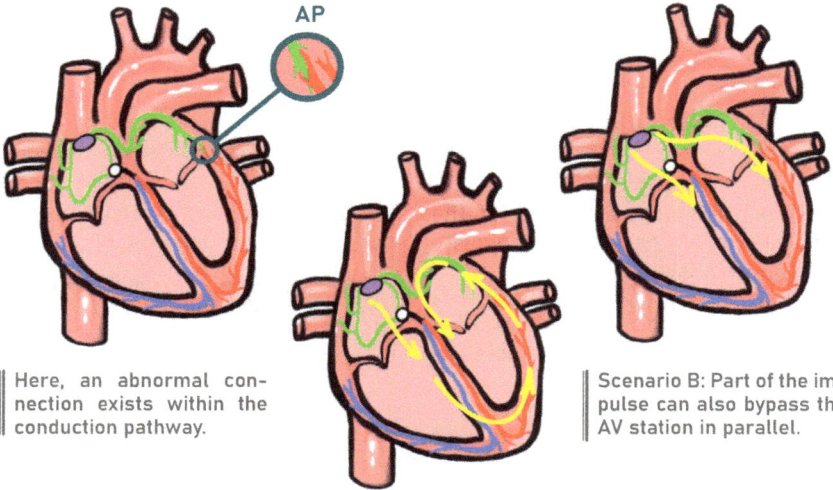

Here, an abnormal connection exists within the conduction pathway.

Scenario A: Reentry circuits can form through the accessory pathway, leading to tachycardia.

Scenario B: Part of the impulse can also bypass the AV station in parallel.

# EVERYTHING AT A GLANCE

**Key Finding** = PR interval under 120 ms!

- In addition, the QRS complex is often wider than 110 ms.

- **Delta waves** are usually visible as small bulges at the beginning of the QRS complex.

- A distinct PR segment is often missing, since the QRS complex starts right after the P wave.

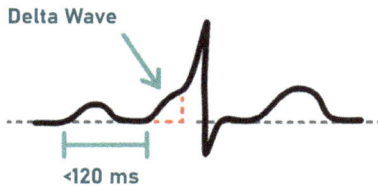

Delta Wave

<120 ms

# BLOCKED GEARS
## AV BLOCK

Our next pathology also involves the PR interval, but here it's the opposite problem: it's too long.

So how can this occur? The AV station normally works with remarkable reliability and precision.

Sometimes, though, something goes wrong with the gears in its mechanism, so it only works partially, or stops working altogether.

In rare cases of a complete shutdown, the auxiliary power plant kicks in, as shown below.

We call this an AV block. It can occur in varying degrees of severity.

> **We distinguish between:**
>
> • **First degree**
> • **Second degree (Mobitz I and II)**
> • **Third degree**

You'll soon see how each degree presents in practice, and we'll take a closer look at each scenario in detail.

Someone here is desperately trying to figure out what's wrong with the gears.

## First Degree

As you might have guessed, the degrees are based on severity. A first-degree AV block is therefore the mildest form and is often symptomless.

In this case, the gears of the AV station are grinding against each other and could use a little fresh oil. The mechanism still turns, but **always too slowly – though at a constant pace.**

You can spot this because the **PR interval is prolonged in every lead, above 200 ms.**

The PR segment represents the time the AV node needs to conduct the impulse from the atria to the ventricles. It's part of the PR interval (P wave + PR segment).

So if the AV node slows down, the segment gets longer, and the PR interval as a whole is prolonged.

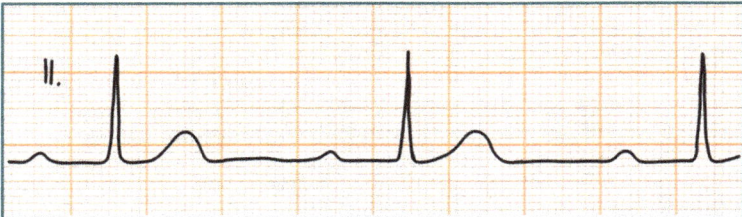

**First Degree:** Anyone who roughly counts the boxes here can see right away that something's off. Even with the naked eye, it's obvious that the AV node is taking far too much time.

## Second Degree

This category actually covers two conditions. A certain Mr. Mobitz generously donated his last name to label them.

Unfortunately, that didn't really do anyone a favor, because it makes mixing them up all the easier.

But don't worry, the Roman numerals I and II help us tell them apart. In fact, we'll put together a neat visual trick in a moment so you won't have to stress about which Mobitz does what.

What they share in common is that they represent a sort of in-between stage: **some impulses are already failing to reach the ventricles, but the atrioventricular transmission isn't completely blocked yet.**

The difference between them lies in how those failures occur.

• **Mobitz I (formerly called Wenckebach):** The AV station turns slower and slower. With each rotation, the gears jam more tightly until a ventricular activation is no longer conducted at all. Then the gears snap back into place, and the whole process starts over.

**In other words, the AV node's ability to pass on an impulse worsens with every cardiac cycle.**

Our gear analogy works nicely here. But you can also picture it as the AV node gradually tiring out. At some point, it becomes so exhausted that it refuses to go on and takes a break. It needs one heartbeat to recover.

At that point, ventricular activation is lost and the QRS complex and T wave disappear entirely.

That heartbeat doesn't happen, and the rhythm "skips." Afterward, the AV node starts over and can again conduct only a limited number of impulses before it needs another brief rest.

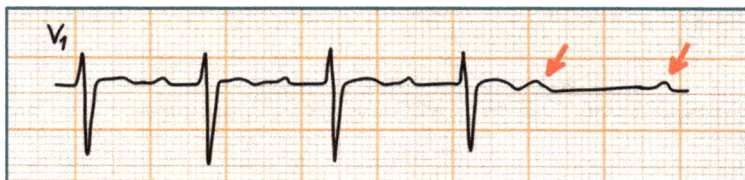

**Second Degree Mobitz I:** What's still a bit tricky to spot in the first two PR segments becomes obvious in the third; it gets progressively longer. The next impulse isn't conducted, and there's complete "silence." As a result, two P waves appear back-to-back (see arrows).

• **Mobitz II:** Unlike before, the AV node doesn't "tire out" here.

On the contrary, it seems to work reliably, with a constant PR interval that doesn't lengthen. But suddenly, a ventricular complex is missing, or sometimes even several in a row.

**Surprisingly, the gears now follow a very specific pattern.** They work flawlessly for a set number of cycles, then - without warning - completely block and prevent a certain number of conductions.

For example, out of three P waves, only the first two are conducted, while the third is not, after which the pattern starts all over again.

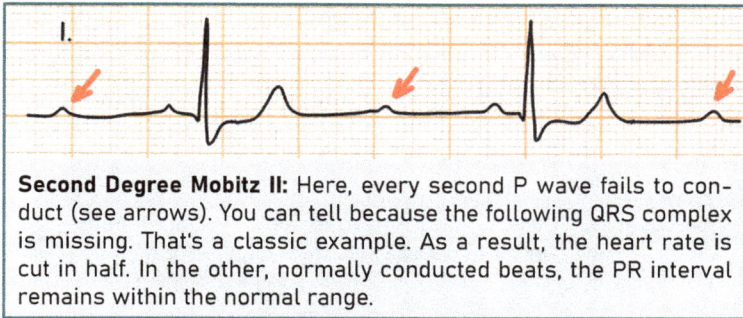

**Second Degree Mobitz II:** Here, every second P wave fails to conduct (see arrows). You can tell because the following QRS complex is missing. That's a classic example. As a result, the heart rate is cut in half. In the other, normally conducted beats, the PR interval remains within the normal range.

**Memory Aid:** Picture the Roman numeral I turned 90 degrees. Now imagine it turning into a tape measure.

The more you pull, the longer the distance it shows. The tape measure stands for the PR segment, which keeps getting longer in Mobitz I.

A tape measure for Mobitz I. By process of elimination, you can already guess what Mobitz II stands for.

## Third Degree

The last one in the group is also the most serious. The gears are completely destroyed, and the mechanism no longer works.

Atrial activation is still driven by the sinus node (P waves), but that's where it ends.

The AV station can no longer pass any impulses from the sinus node down to the ventricles.

The ventricles now have a kind of "will of their own," since they can't follow the rhythm set by the sinus node. Instead, they're powered by their own independent source, forming a ventricular escape rhythm. This usually originates from the auxiliary power plant of the AV station, sometimes from the His bundle.

**That's why a third-degree AV block is recognized by P waves and QRS complexes occurring completely independently of each other.**

The escape rhythm typically runs at well below 60 beats per minute, which is another key clue.

Third degree means absolute chaos, as you'll see on the next ECG.

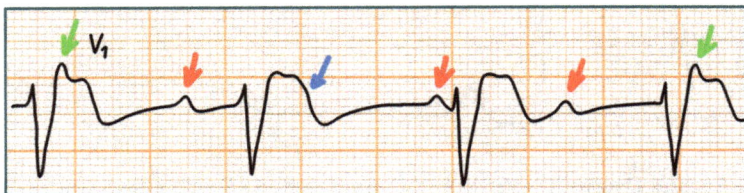

**Third Degree:** A typical feature here is the dissociation between P waves (arrows) and QRS complexes. Meaning they're no longer linked.

It can even go so far that some P waves fall right in the middle of a QRS complex (green arrows). In one case, the P wave is completely swallowed by the T wave, leaving nothing but a small bulge at the edge (blue arrow).

Another striking feature is that the ventricular escape rhythm produces wide QRS complexes that lie so close to the T wave they can blend directly into it. This creates the tall "mountain" you see after each QRS complex.

What's also clear is the slower rate of the escape rhythm compared to the atrial rate. We see six P waves, but only four ventricular complexes.

**Tip!**
If you notice a low heart rate along with unusually shaped QRS complexes, the first thing to do is look for P waves. Specifically, keep an eye out for small deflections appearing at regular intervals.

It's suspicious if they don't show up in their usual spot before the QRS complex. Don't forget: they might also be hiding right inside the QRS complex.

As you can see above, the deflections marked by the arrows are all spaced about equally apart - so they're most likely the "missing" P waves.

If you spot deflections like these and their frequency is higher than that of the QRS complexes, you're dealing with a third-degree AV block.

## Nice to Know – Clinical Picture of the Different Degrees

As mentioned earlier, a first-degree block usually causes no symptoms and therefore generally requires no treatment.

Second-degree Mobitz I is also often asymptomatic, though it can sometimes cause reduced exercise capacity. If that happens, treatment with medications that improve cardiac performance may be considered.

Mobitz II is typically noticeable to patients and has a greater impact on performance. In many cases, a pacemaker is already indicated, since the risk of sudden progression to a third-degree block is high.

A complete block must always be treated with a pacemaker to ensure a physiological heart rate and to bring the atria and ventricles back into sync.

# EVERYTHING AT A GLANCE

**First Degree:** The PR interval is uniformly prolonged (>200 ms), but no beats are lost.

**Second Degree:**
- **Mobitz I:** The PR interval gets longer with each beat until a ventricular activation fails. Then the cycle starts all over again.

- **Mobitz II:** The PR interval stays constant and is not prolonged. However, conduction through the AV node drops out in regular patterns, for example, every second beat. (**Careful!** Mobitz I can also form patterns. The key difference is in the PR interval!)

**Third Degree:** The AV node has completely failed. You can see P waves that are no longer linked to the ventricular complexes (**dissociation**). Wide, deformed ventricular complexes from the escape rhythm appear. The rate of the P waves is higher than that of the QRS complexes.

# BROKEN CABLE
## BUNDLE BRANCH BLOCK

Now things are getting interesting! For the first time on our journey through the mountains and valleys of the cardiac cycle, we'll specifically need the precordial leads.

This time it's especially important to distinguish the right ventricle from the left in order to make an accurate diagnosis.

We've finally left the PR segment behind and reached the next stage: the QRS complex. In other words, it's all about ventricular depolarization. More specifically, this chapter deals with what we call bundle branch blocks.

## So, What's a Bundle Branch Block?

As the name suggests, this means that one or both of the bundle branches, which normally activate the ventricles separately, are partially or completely blocked.

Most of the time, however, only one of them is affected.

As you can see in the illustration, the problem isn't with the AV "charging station" but lies deeper within the conduction system.

A sneaky critter has actually chewed through one of the cables, in this case, the left one. Because of that, the activation of this ventricle has to take a detour.

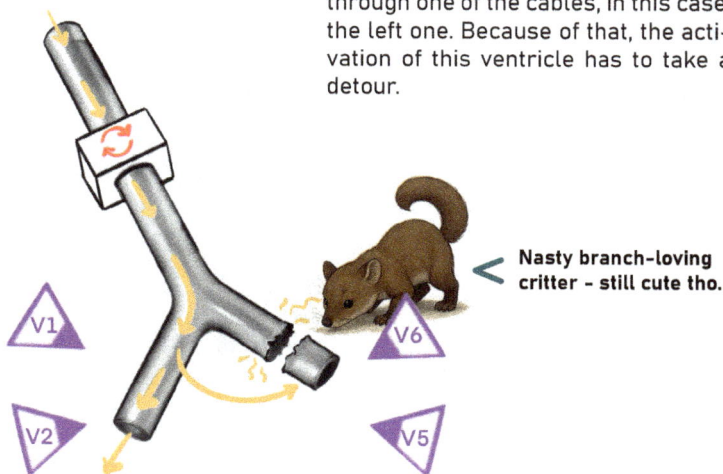

Nasty branch-loving critter - still cute tho.

Apparently, bundle branches are one heck of a delicacy.

Think of it like this: part of the impulse from the healthy ventricle branches off and crosses the septum to reach the affected side.

That way, the ventricle on that side still gets activated even though its usual "cable" is cut.

The catch is that the whole ventricular depolarization now takes longer than usual. Our "hearing aids" (the leads) pick up on this delay, and that's the key sign of a bundle branch block.

If the QRS duration is longer than 120 ms, our alarm bells should be ringing.

All that's left now is to figure out which side is blocked.

How to do that will be explained in just a moment. Take another look at the precordial leads and recall which ones specifically reflect the right and left ventricles.

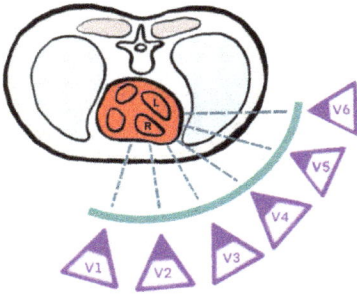

Refresher: Assigning the Precordial Leads

**Important!**

For the sake of simplicity, there's one thing we haven't mentioned yet. The left bundle branch actually splits again inside the ventricle. It divides into an anterior and a posterior branch (in fancy terms: fascicles). Why? Well, the left ventricle simply has more muscle mass, and this split helps distribute the charge more evenly.

And as you might have guessed: sometimes only one of these two branches gets blocked. That's what we call a fascicular block.

We classify them as follows:

- **Left bundle branch block (both fascicles involved)**
- **Left anterior fascicular block**
- **Left posterior fascicular block**
- **Right bundle branch block**

For you, as a student, the main thing is to at least be able to tell a complete left bundle branch block from a right bundle branch block. That's where we'll keep our focus.

Spotting a left fascicular block isn't that easy. It takes more experience and is therefore more of an advanced topic. Plus, fascicular blocks are clinically less relevant anyway.

## Left Bundle Branch Block

The illustration at the beginning of this chapter shows a left bundle branch block. It also marks how the impulse takes its detour across the septum to reach the left ventricle. And that's exactly what we can see on the ECG.

**We observe part of the charge moving away from the right ventricle (V1, V2) and heading toward the left (V5, V6).**

As already mentioned back in the chapter "The System", when the electrical charge moves away from a lead, it creates a negative deflection on the ECG, and a positive deflection, when it moves toward it.

That's why, in a left bundle branch block, **you'll mainly see a strong negative deflection in V1 and V2:** the charge is moving away from the right ventricle.

**In V5/V6, we find a broad, bulky, notched positive QRS complex,** usually without any negative components. Here we're watching the charge arrive on the left side, often in several small bursts, which explains the little extra waves "on top."

So, for the diagnosis, just follow the path of the charge: wherever it's headed, that's where the block is!

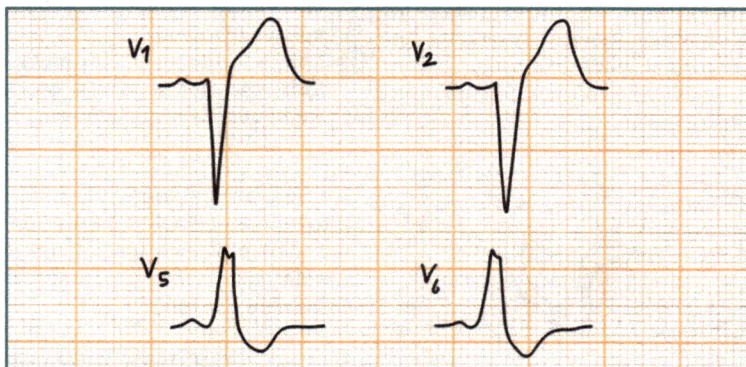

**Left bundle branch block:** The first thing you'll notice is the prolonged QRS duration, which is typically a bit more pronounced on the affected side. In this case, it's about 140 ms. You can clearly see the deep negative waves in V1 and V2. Leads V5 and V6 show a textbook picture: bulky, broad QRS complexes with little notches on top.

## Right Bundle Branch Block

As you'd expect, it behaves pretty much the opposite of a left bundle branch block. The signs of the charge shifting are just a bit more subtle. What stays the same, of course, is the prolonged QRS duration. But we don't see a strong negativity in V5/V6, because the left, more powerful ventricle has much less trouble passing on part of its charge.

Still, there is a telltale clue of the charge moving away from the left ventricle: the so-called **S persistence.**

This means you can see S waves in V5 and V6, that is, negative waves at the end of the QRS complex. In healthy individuals, that wouldn't be the case. Normally V4 is the last lead to still show an S wave.

On top of that, **you'll find a broad, notched QRS complex in V1/V2,** created because the charge is now heading into the right ventricle. In right bundle branch block, this usually shows up as the **classic M-shape.**

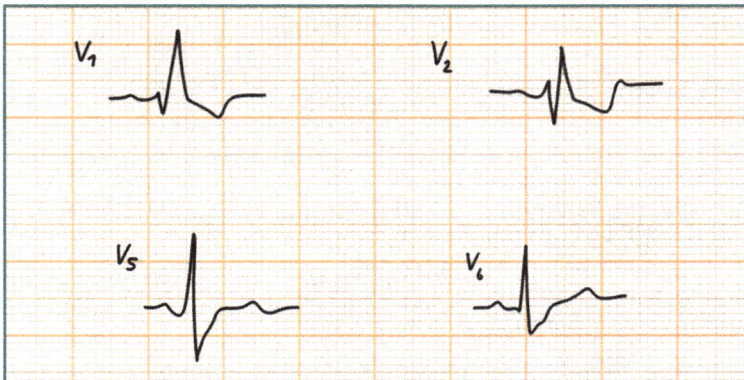

**Right bundle branch block:** When we look at the right ventricle, meaning V1/V2, the M-shaped QRS complex immediately stands out. In V1 it's only hinted at, but in V2 it's already clearly visible. That's definitely not what a normal ventricular complex looks like. This is what we call an rSR' complex, as already described back in the chapter "Cardiac Cycle".

Naturally, the QRS duration is also clearly prolonged, here about 160 ms. In V5/V6 we see S waves (S persistence), caused by the delayed, non-physiological shift of charge to the right, making the charge move away from V5/V6.

## Incomplete Bundle Branch Block

When the **QRS duration is only between 110 and 120 ms** - so just slightly prolonged - we call it an incomplete bundle branch block. Depending on how it shows up, it usually produces the same QRS shapes as a complete block, just a bit less pronounced. That's why you need to look a little more closely whenever the QRS duration drifts just outside the normal range.

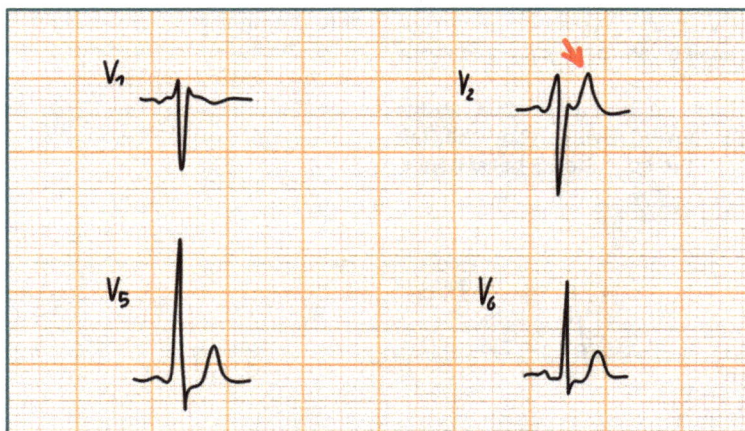

**Incomplete right bundle branch block:** At first glance, this incomplete right bundle branch block has pretty much all the features of a complete block when it comes to shape. In V1 we can just about make out an M-shape (with a bit of imagination, since the positive waves are very small), and in V5/V6 we see S persistence.

When we measure the QRS duration precisely with our computer tool, we get 115 ms. That's above our cutoff of 110 ms, but not enough to diagnose a complete right bundle branch block.

**Careful, pitfall!** In V2, what you see at the end is already the T wave (arrow), not part of the M-shaped QRS complex.

**Important!**
If a bundle branch block is present, the ST segment can't be evaluated for other questions anymore. That's because it reflects repolarization, which also gets distorted by the abnormal shift of charge caused by the block. In other words, the ST segment loses its diagnostic value.

This can make diagnosing a heart attack by ECG very tricky. If a block is present, you'll have to rely on other diagnostic methods when a heart attack is suspected.

Why the ST segment plays such a central role in diagnosing an infarction is something you'll discover in the corresponding chapter.

## Nice to Know – Causes and Symptoms

A bundle branch block is always the result of damage to the conduction tissue. In younger people this can sometimes be due to a congenital malformation, but in older folks much more often the trigger is ischemia, myocarditis, or increased intraventricular pressure (for example, from a valve stenosis). In that sense, it can also serve as a clue to a past cardiac event.

Because of the abnormal spread of the electrical charge, the two ventricles no longer contract in sync. Interestingly, most patients don't actually experience any symptoms. In the case of a left bundle branch block, however, the asynchronous contraction of the two ventricles can reduce the pumping performance of the left ventricle, which may show up as reduced exercise capacity. In such cases, a pacemaker can be used to resynchronize the ventricles.

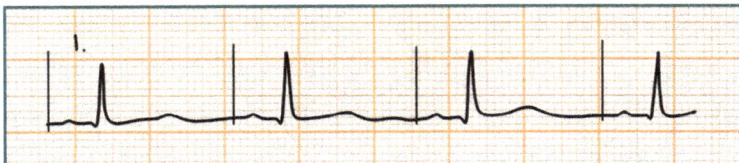

**What's that?** Notice how regularly these little lines show up, and always in the exact same spot. So they can't be artifacts (like from movement).

What you're actually seeing here is a pacemaker on the ECG. Each heartbeat is being triggered by one of those spike-shaped impulses. In some patients, though, they only appear every now and then - for example, if the heart occasionally skips a beat and the pacemaker just steps in when needed.

Keep in mind that these pacemaker spikes can sometimes be smaller. In many cases, though, they're quite tall and easy to spot.

# EVERYTHING AT A GLANCE

**Left Bundle Branch Block**
- Deep S waves in V1 and V2
- Broad, bulky positive QRS, often with notches on top, in V5 and V6
- QRS duration >120 ms

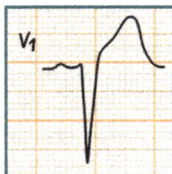

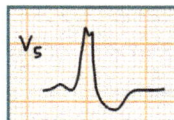

**Right Bundle Branch Block**
- M-shaped QRS in V1 and V2
- S persistence (S waves present) in V5 and V6
- QRS duration >120 ms

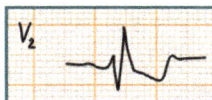

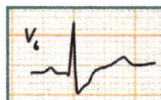

**Incomplete Bundle Branch Block (left or right):**
- Same morphological changes, but often less pronounced
- QRS duration between 110 and 120 ms

**Careful:** The ST segment can no longer be used diagnostically, since a bundle branch block makes it lose its value!

# THE ISCHEMIA WRECKING BALL
## MYOCARDIAL INFARCTION

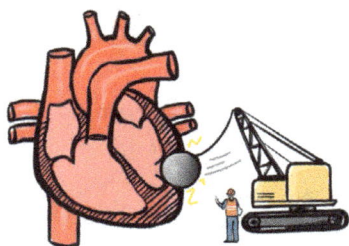

**What do wrecking balls have in common with heart attacks? Right, absolutely nothing!**

And that's exactly what makes them perfect for creating a few great memory aids and turning them into a central part of this chapter.

## What Does Myocardial Infarction Mean?

When we talk about a myocardial infarction, we're always referring to part of the heart wall being affected by ischemia (= lack of oxygen) and becoming damaged as a result.

Depending on how long the ischemia lasts, the damage may be reversible, or it may become permanent. In most cases, at least part of the affected area ends up with permanent injury.

These acute injuries to the myocardium alter the way the electrical charge spreads in that region.

As a result, we find typical signs in the specific leads that are closest to the area where the problem is located.

> We look out for three things:
>
> - **ST segment changes (elevations, depressions)**
> - **Negative T waves**
> - **Pathological Q waves**

There are other signs that can point to ischemia, but they're often not clear-cut and require much more experience to interpret correctly. That's why we'll focus on the straightforward ones, so you'll be able to spot them with confidence.

We'll first look separately at pathological Q waves, ST changes, and T waves. After that, you'll see from real-life examples how these findings come together to make the right diagnosis.

## Pathological Q Waves

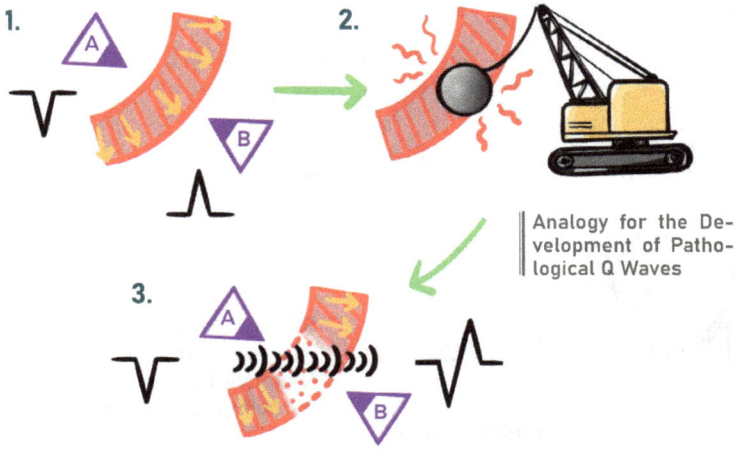

Analogy for the Development of Pathological Q Waves

**1.** You're looking at a healthy heart wall with good blood supply. To understand the figure, you also need to know that excitation in the heart wall always spreads from the inside out.

So if we imagine a fictional lead on the inside of the heart - labeled "A" in this case - it would "hear" the charge moving away from it and therefore record a negative deflection.

Another fictional lead sits on the opposite side, on the outside. This one "hears" the charge coming toward it.

**2.** But there it happens: a thrombus blocks off a branch of the coronary arteries, and part of the heart wall gets slammed by the ischemia wrecking ball. **If the ischemic state lasts too long, a necrotic area usually develops.**

This region (shaded) is so damaged by the lack of oxygen that no electrical charge moves through it anymore.

**3.** Our "hearing aid" on the outside uses this area like an open window through which the sounds from inside can escape. This way, it's actually possible to "hear" what's going on within. As a result, some of the negative components from the inside get mixed in with the positive ones on the outside. This is exactly what happens in our actual ECG leads.

**The negative components sneak their way into the QRS complex, right at the spot where the Q wave sits. That makes it even more negative.** In fact, this can even cause a "fake" Q wave to show up in leads that normally wouldn't have one at all (V1–V4).

The closer a lead is to the infarct area, the stronger the negative deflection of the Q wave becomes. If it lies directly over the infarct, you may sometimes see not a QRS complex at all but just **one single, large, purely negative wave: a so-called QS complex.** This happens because the negative components are so strong they essentially "swallow up" the rest of the QRS complex.

Q waves are considered pathological if **any** of the following apply:

- **In V1–V4 they're always problematic, no matter how small they are**

- **In all other leads, if they are at least 1/4 the amplitude of the following R wave**

- **If they last longer than 30 ms**

## 4 4 3

| Precordial up to V4 | 1/4 of R | longer than 30 ms |

So many damn rules – it's pure lunacy! I'll just cut the crap and remember 4-4-3.

### Important!
A Q wave is only truly a Q wave if there's no positive wave before it in the QRS complex, even if that positive wave is tiny. Otherwise, it's an S wave.

Since necrotic areas usually develop only after the acute phase of a myocardial infarction, pathological Q waves are generally a sign of an older infarction. Under certain circumstances, however, they can also appear during the acute stage.

A persistent pathological Q wave will in many cases remain for life as a marker of an old infarction.

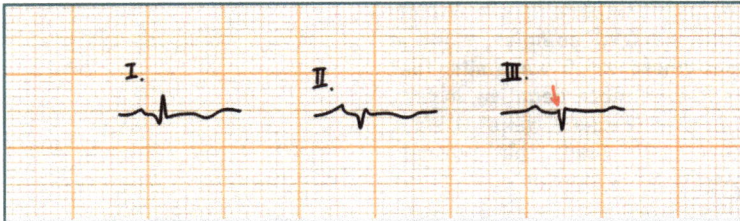

**Pathological Q Wave: Yes or No?** As you can see here, you really have to look closely to decide whether you're dealing with a pathological Q wave or not. In the first lead it's obvious: it's about one box wide (>30 ms) and also larger than 1/4 of the following R wave. The second lead also meets the 30 ms criterion.

Even though the third case looks just as clear at first glance, you have to be careful here. The tiny little notch at the beginning (arrow) is actually an R (because it's positive)! No matter how small a positive wave at the start of the QRS complex may be: it means that the negative wave that follows is not a Q but an S!

# A Small Favor

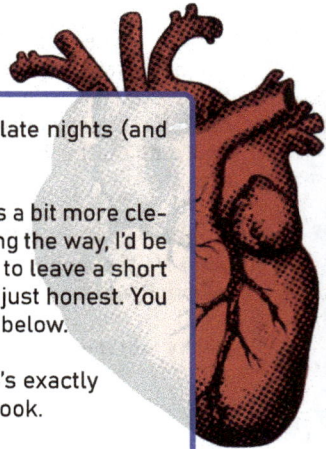

This book started as a small idea, and many late nights (and coffees) later, it finally became real.

If you liked it so far and it helped you see ECGs a bit more clearly, or just made you smile once or twice along the way, I'd be super grateful if you could take only a minute to leave a short review on Amazon. It doesn't have to be long, just honest. You can do so instantly, by scanning the QR Code below.

Your words might be helpful to someone who's exactly where you were when you first opened this book.

And they remind me that all those late nights were totally worth it.

**Thank you so much for your support!**

**Philipp**

## ST Elevations and T Wave Inversions

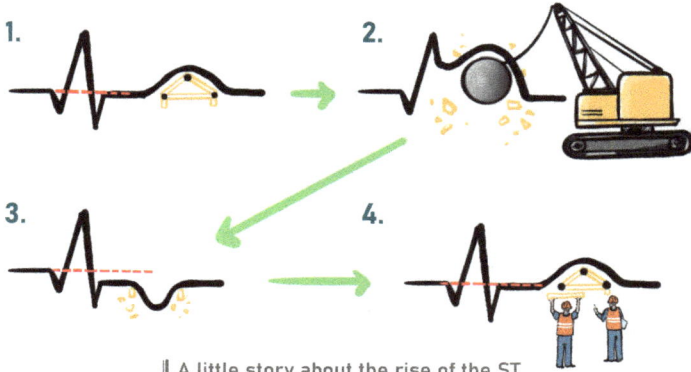

A little story about the rise of the ST segment and the fall of the T wave.

**1.** Imagine that the material making up the T wave is rather unstable. On its own, it couldn't keep its hilly shape. That's why it's supported by a scaffold. The ST segment is flat and lies at the same level as the PR segment.

Together, the ST segment and the T wave represent the phase of repolarization.

**2.** Once again the crane rolls in with its massive ischemia wrecking ball. **The lack of oxygen disrupts repolarization in the infarct area.** The ball smashes into the supporting structure of the T wave and destroys it. With so much force behind it, the wrecking ball even **yanks the ST segment upward. Now it's no longer at the same level as the PR segment.**

You'll see this mainly in the acute phase. An ST elevation appears shortly after the onset of a heart attack and

can last for a few hours up to a few days. After that, the ST segment transitions into the next phase..

**3.** After some time, the ECG changes further. At this stage, the wrecking ball has already been removed, and what's left behind is the rubble. **Since the material of the T wave is so unstable, it now droops downward.**

It usually tapers to a rounded point and is therefore described as **V-shaped.** In ischemia it's typically symmetrical, meaning the descending and ascending limbs look similar. In contrast, non-symmetrical negative T waves are usually nonspecific.

**An important point: additional ST segment changes are often visible here,** most commonly a slight ST depression, as shown in the figure.

An isolated negative T wave without accompanying ST changes, on the other hand, can point to a past ischemia or another cardiac cause.

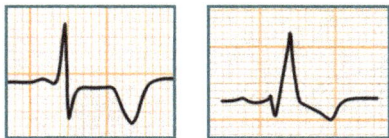

On the left: a symmetrical T wave inversion (V-shaped) with ST segment depression. On the right: a non-symmetrical T wave inversion, which does not indicate ischemia.

**4.** After about four weeks, the repair work on the T wave by our diligent repair crew is nearly finished, and a temporary scaffold is back in place. By this stage, the ST segment has returned to its usual appearance. However, it's also possible that the negative T remains after an infarction and never resolves.

As a beginner, you don't need to be able to pinpoint the exact stage of an infarction on the ECG. The take-home message is that the different stages blend into each other, and that's why you should be on the lookout for varying degrees of these signs.

### Important!

If a myocardial infarction is present, ST elevations may occur, but that's not always the case. When they are visible, we call it a **STEMI** (ST elevation myocardial infarction). If they're absent, it's an **NSTEMI** (Non-ST-elevation myocardial infarction).

But an NSTEMI is not taken any more lightly than a STEMI. The difference isn't only the absence of ST elevations, but also the underlying pathophysiology. While a STEMI usually results from a complete coronary occlusion, an NSTEMI is more often caused by a partial blockage.

On the ECG, an NSTEMI often shows other signs of ischemia, particularly ST depressions and T wave inversions. And additionally, for diagnosing an NSTEMI, blood markers - especially cardiac enzymes - play a central role.

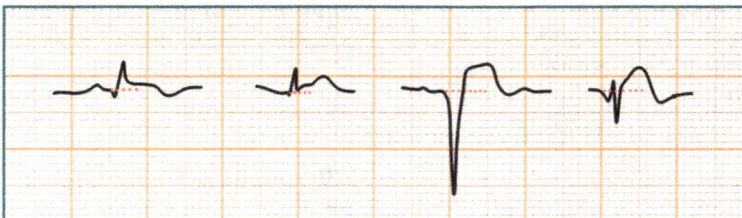

**Varying degrees of ST elevation:** In theory, ST elevations smaller than 0.1 mV are insignificant. However, as a beginner it's best to treat every elevation as potentially pathological.

**But watch out!** It only counts as an elevation if the ST segment is higher than the PR segment right from the start. If it begins at PR level and only rises upward from there, that does not qualify as ST elevation! In the figure, the second elevation just barely meets this criterion.

## Physiological T Wave Inversion

Not every negative T wave is pathological.

**Here's a simple rule of thumb to remember:**

Negative T waves in the leads that are located to the right (from the patient's perspective) of the sagittal midline can be harmless.

That would be **aVR, V1, and III.**

⚠ **Important!**
This only applies to the latter if the QRS axis is rotated far to the left. **Otherwise, a negative T in lead III is not normal.**

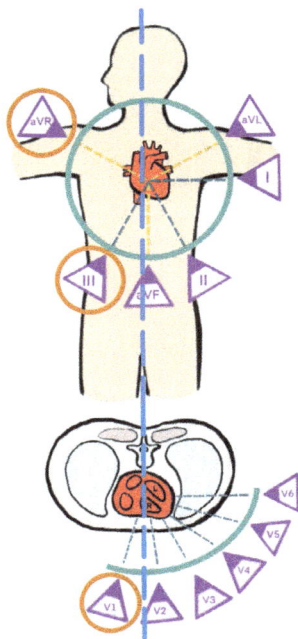

In these three leads, negative T waves can be physiological.

## ST Depressions

You'll need just a bit more information to interpret ST depressions correctly.

> We divide them into three groups based on their shape:
>
> - Upsloping
> - Downsloping
> - Horizontal

**Horizontal and downsloping ST depressions point to an acute ischemia.** Upsloping ST depressions, on the other hand, do not. This is usually physiological and often appears with tachycardia or hyperventilation.

- **Upsloping ST depressions** are already below the PR level right after the QRS complex (red dashed line) and then rise back up over the course of the ST segment.

- **Downsloping ST depressions** start just slightly below the PR level but then drop further and further down.

- **Horizontal ST depressions** remain at the same low level throughout the entire ST segment until the T wave begins.

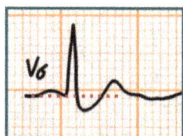 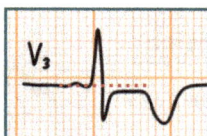 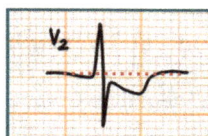

‖ Upsloping ‖ Horizontal ‖ Downsloping

**Tip!**
If you have trouble remembering this, just think about what a submarine would do if it ran into a **storm (ischemia).**

It would **dive down** to get below the big waves, then continue moving forward **horizontally** until the storm passes. Rising back up wouldn't make any sense as long as the bad weather lasts.

**Exception:** There is one type of upsloping ST depression that, when combined with a very tall T wave, does indicate ischemia. It's usually seen in the precordial leads and is called a **De Winter T wave.**

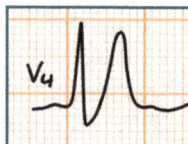

‖ De Winter T wave

## Infarct Localization

To localize an infarct area, you need to find signs **in at least two anatomically neighboring leads.** For example, in III and aVF or in V4 and V5. **Only then do they carry real diagnostic weight!**

This way, you can gather the first clues about the approximate site of the infarct. We refer to the specific wall that's affected.

To help you understand how the different leads guide us in locating the infarct area, they're all shown here once again. The septum and the anterior, lateral, and posterior walls are color-coded.

With this illustration, you can now easily follow how the table of resulting infarct localizations is put together.

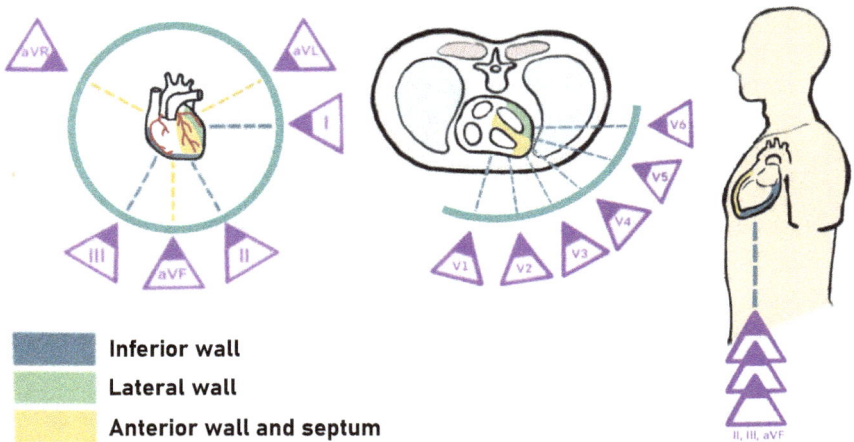

- Inferior wall
- Lateral wall
- Anterior wall and septum

| V1 and V2 | Septal Infarction |
|-----------|-------------------|
| V3 and V4 | Anterior Infarction |
| V5, V6, I, aVL | Lateral Infarction |
| II, III, aVF | Inferior Infarction |

**Mixed forms are possible too!**

## Nice to Know – Generalized ST Elevations

If you find ST elevations in (almost) all leads, it's not a heart attack affecting the entire muscle but much more likely an acute pericarditis. The same can also happen with myocarditis if the pericardium is involved as well. Don't let that throw you off!

### Tip!

A heart attack is hardly ever an incidental finding. Most of the time, typical symptoms stand out: weakness, shortness of breath, angina pectoris, chest pain with radiation, or even nausea and upper abdominal pain. The last two especially in women.

On top of that, there are often clues from the patient's history, such as a collapse, previous heart attacks, or risk factors. So it's always important to consider the overall picture.

### Important!

Because of the anatomy, a myocardial infarction almost always affects the left ventricle. It's much more muscular, which is why most branches of the coronary arteries - including the terminal branch of the right coronary artery - supply the left ventricle and the septum.

An infarction of the right ventricular wall can occur, but it's rare. **That's why terms like anterior, lateral, and posterior infarction usually refer to the left ventricle.**

Even though it may look as if V1 and V2 are observing the right ventricular wall, if we see changes there it's far more likely to be an infarction of the underlying septum.

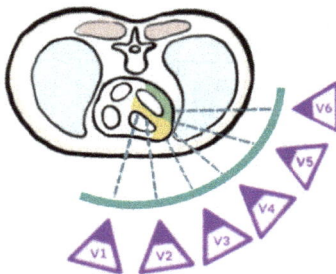

V1 and V2 usually look straight through the outer wall of the right ventricle and focus more on the septum.

V3 can also be affected in pronounced septal infarctions, but it's more commonly associated with an anterior infarction of the left ventricle.

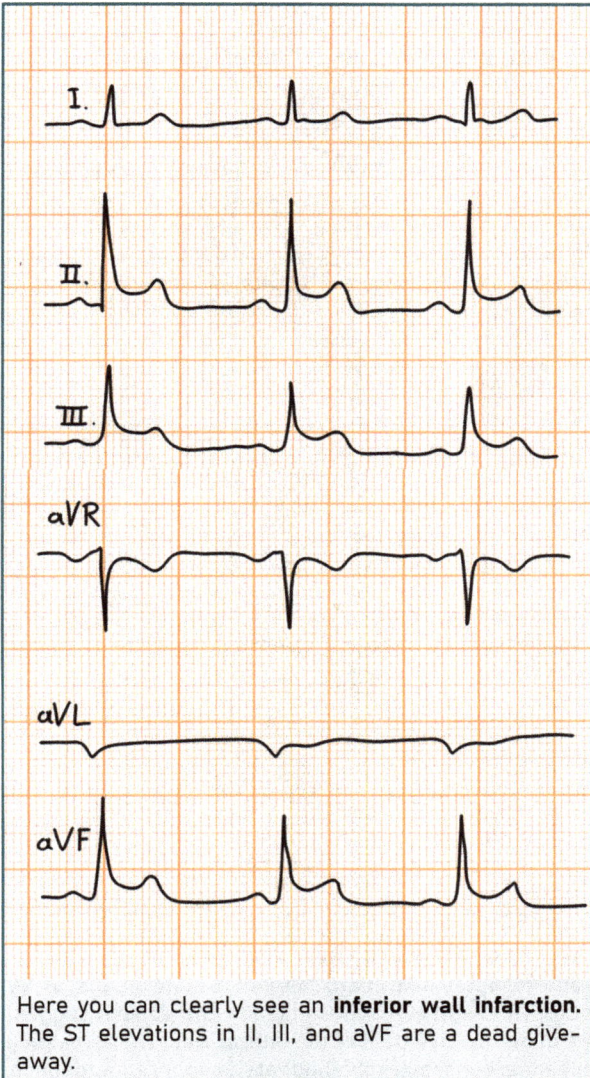

Here you can clearly see an **inferior wall infarction**. The ST elevations in II, III, and aVF are a dead give-away.

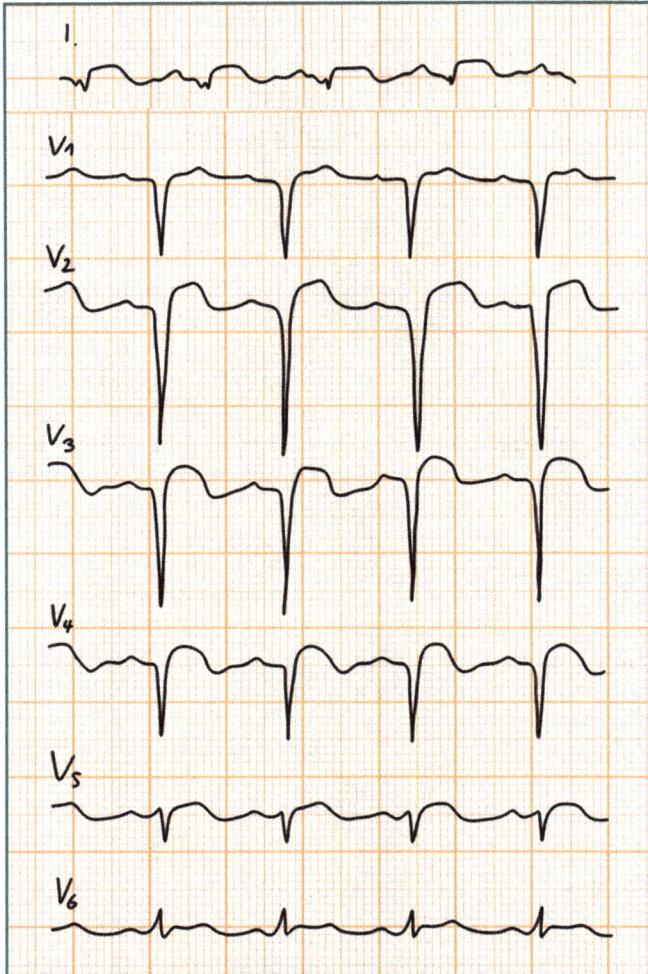

**An anteroseptal infarction.** Marked ST elevations in V2 through V4. Slight elevations in lead I and V5 also suggest some minor involvement of the lateral wall. The remaining limb leads are unremarkable and not shown here due to space.

As a little bonus, you may have noticed that the transition occurs way too late, between V5 and V6.

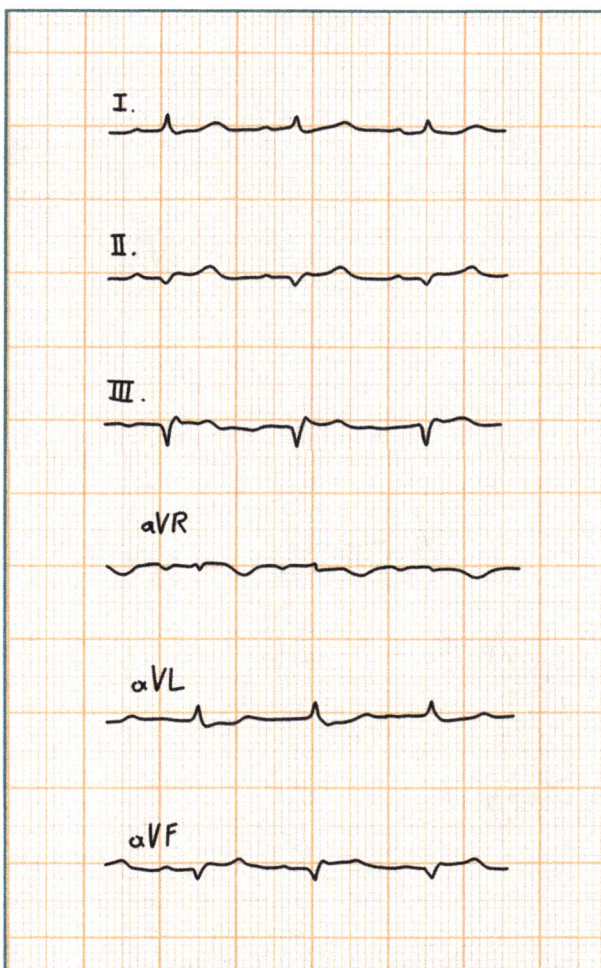

What looks like an ECG at a different scale is actually from someone whose cardiac excitation shows low voltage. That's why the QRS complexes look a bit smaller.

What really matters here, though, are the pathological Q waves in II, III, and aVF. If you look closely, you can also spot a slight ST elevation in these leads. So once again, we're dealing with an **inferior infarction.**

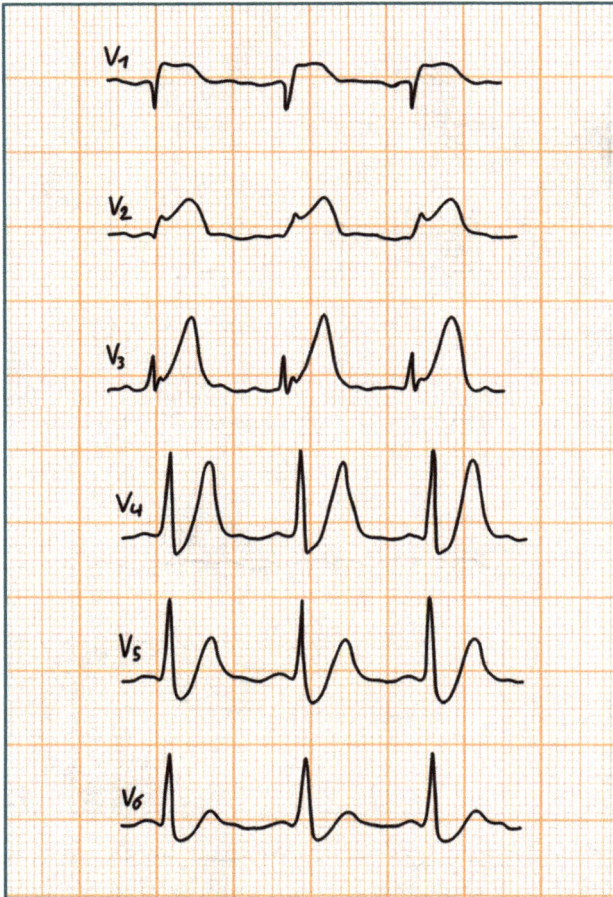

Finally, here's a typical **anteroseptal infarction.** The ST elevations are mainly concentrated in V1 and V2 (septum). Lead V3 is only mildly affected, and V4 shows no abnormalities.

In addition, we find De Winter T waves in the precordial leads V3 and V4, and to a lesser degree in V5.

**Important!**
Final message: Even a completely unremarkable ECG does not rule out an infarction with 100 percent certainty! Never rely on it alone.

When typical symptoms are present, always play it safe!

# EVERYTHING AT A GLANCE

**NSTEMI:**
No ST elevations. The key findings here are pathological Q waves, ST depressions, and inverted T waves (V-shaped).

**STEMI:**
ST elevations/depressions, pathological Q waves. Inverted V-shaped T waves can also be indicative.

**Localisation:**
**Signs in at least two neighboring leads!**

| | |
|---|---|
| **V1 and V2** | Septal Infarction |
| **V3 and V4** | Anterior Infarction |
| **V5, V6, I, aVL** | Lateral Infarction |
| **II, III, aVF** | Inferior Infarction |

**Signs:**
**Q waves** are pathological
- if they occur in V1 through V4,
- if they last longer than 30 ms,
- or if they are larger than 1/4 of the following R wave.
**Remember 4-4-3!**

**ST elevations** can take different shapes. Even if the ST segment lies just slightly above the PR segment, it should be considered potentially pathological.

**Downsloping and horizontal ST depressions** point to ischemia, while upsloping ones do not. **Exception:** an upsloping depression with a tall, peaked T wave (De Winter T wave) in the precordial leads.

**V-shaped negative T waves** only indicate an acute infarction when combined with ST depression, otherwise they more likely point to a previous infarction.

Negative T waves can be physiological in aVR and V1 (and also in III if the QRS axis is markedly leftward).

# CONFUSION ALERT!
## CALCIUM AND POTASSIUM

It's easy to get mixed up about what potassium and calcium actually do to the heart.

But don't worry, by the end of this chapter, you'll have it all sorted out.

### Why Are They So Important?

To answer that, we need to take a quick detour into physiology. Don't fear though, we'll keep it short.

For the pacemaker cells in the sinus power plant to generate impulses, they need calcium, potassium, and sodium.

The figure below gives you an overview of how these three work together to generate an impulse:

**1.** A slow influx of sodium starts to raise the voltage until it reaches a certain threshold (red dashed line).

**2.** Once that point is reached, a large number of channels open, allowing calcium to rush into the cell. This influx generates the impulse.

**3.** At the peak, a strong efflux of potassium takes place, bringing the cell back down to its original voltage level.

After that, the whole cycle starts over again.

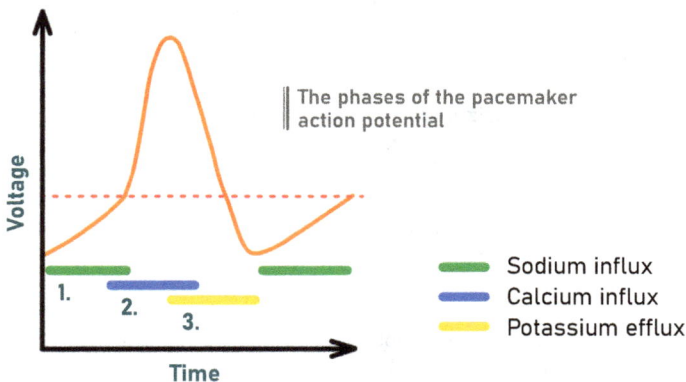

The phases of the pacemaker action potential

Sodium influx
Calcium influx
Potassium efflux

**Notice how the phases overlap.** A small amount of calcium is already flowing in before the threshold is reached and the big influx begins. Likewise, at the end of the rise, potassium is already starting to flow out before all the calcium channels have closed.

> **Important!**
> Sodium only acts as a kind of kickstarter for the cell. That's why hypo- or hypernatremia usually doesn't have a direct effect on pacemaker cell activity, and you won't see any changes on the ECG.
>
> So instead, we'll focus on the two big players: calcium and potassium.

## Electrolyte Disorder – Isn't That What the Lab Is For?

The point of this chapter isn't to teach you how to diagnose an electrolyte disorder based solely on the ECG. That's obviously what a blood test is for.

But if the labs show that something's off with the electrolytes, the ECG can tell you whether the imbalance is already affecting the heart. If the changes are clearly visible, the risk of arrhythmia is high and action needs to be taken.

So, in a way, the ECG helps you gauge how severe an electrolyte disorder really is.

## Calcium

While mild calcium disturbances often cause no visible changes on the ECG, severe hypo- or hypercalcemia can have significant cardiac effects.

That said, compared to potassium, calcium usually has a much less dramatic impact on the heart.

Calcium tends to affect the gastrointestinal tract (e.g., nausea, vomiting, ileus) as well as the central and peripheral nervous system (e.g., seizures, coma, tetany) more strongly.

> With **pronounced calcium disturbances**, the ECG shows the following:
>
> **Hypercalcemia**
> - Shortened QT interval
>
> **Hypocalcemia**
> - Prolonged QT interval

> **Tip!**
> When it comes to arrhythmias, just remember: the **"C" in calcium stands for calm**. It doesn't cause nearly as much chaos in the heart as potassium does.
>
> That said, calcium disorders are still nothing to take lightly because of their effects in other systems. But as a mnemonic, it's a neat way to keep them apart.

## Visualizing the Effects of a Calcium Disturbance

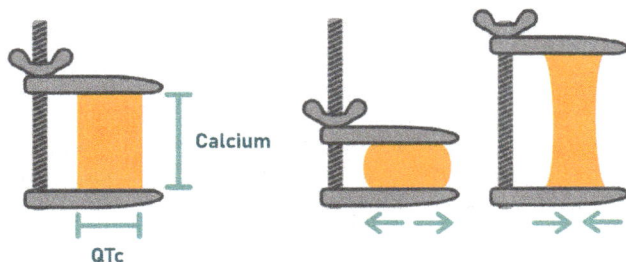

Calcium

QTc

The height of the rubber block represents the calcium balance, while its width represents the QTc interval.

**Picture it like this:** What you see above is a block of sticky rubber held in place by a screw clamp.

Since the clamp itself kind of resembles a "C," it's an easy way to remember that these are the effects of calcium. At the same time it also stands for the c in QTc, so you know which value is being affected.

The QTc interval is represented by the width of the rubber block. The height of the screw clamp's surfaces symbolizes the amount of calcium in the blood.

If calcium drops, the clamp tightens and compresses the block. As a re-

sult, it bulges outward to the left and right, making it wide, so the QT interval is prolonged.

If calcium is too high, the clamp widens. Because the block is sticky and adheres to the clamp surfaces it is pulled apart and becomes thinner. Therefore the QT interval shortens.

**For cardiac function, the main concern is a prolonged QT interval, as it significantly increases the risk of arrhythmias.** To estimate it, you can use the halving rule we already discussed in the "Report" chapter.

Or just remember the cutoff value for the corrected QT interval (QTc):

> **QTc $\leq$ 450 ms**

**Important!**
A prolonged QTc interval can also be caused by various medications.

**As a simple rule of thumb, remember that many of the drug classes starting with "anti-" can do this:**

antidepressants, antipsychotics, antibiotics, antiemetics, antiarrhythmics, and so on.

## Potassium

Potassium has a much more dramatic effect on heart rhythm. **Too much of it in the blood massively slows down conduction.** The more potassium, the slower it gets. You'll notice this mainly because the QRS complex becomes wider and wider.

That makes sense - if the charge moves more slowly through the ventricles, it takes longer until they're fully depolarized. As a result, **the QRS complex stretches out more and more.** At some point it becomes so wide that it merges with the T wave, forming a **sine wave-like complex.**

It's called that because it resembles a mathematical sine wave. It looks so unhealthy that even beginners will immediately spot it.

And that's a good thing, because the next step after the sine wave complex is, in many cases, already **ventricular fibrillation or asystole!** At this point, action must be taken immediately.

Next, we'll look at hyperkalemia and hypokalemia separately.

## Hyperkalemia

- T waves tall and peaked

- P wave flattening

- QRS duration prolonged

**In extreme cases: sine wave-like cardiac activity (ST merging due to QRS prolongation)**

The signs of hyperkalemia usually appear in roughly the order listed above. The first are tall, peaked T waves with flattening of the P wave, along with a widening QRS that becomes more and more dramatic, until the complex eventually merges with the T wave to form a sine wave-like pattern.

All in all, there's a lot going on here. And it's something you'll definitely want to remember. The figure on the next page will help you with that.

## How to Remember the Effects of Hyperkalemia

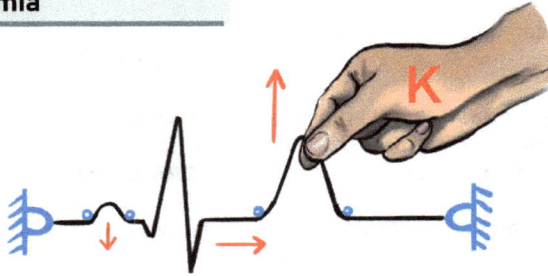

The hand represents potassium. The higher the level in the blood, the higher the hand rises.

Imagine this ECG as a cord that can't be stretched. The blue structures are all fixed in place and don't move.

As potassium levels rise, so does the hand pulling on the cord of the T wave, making it taller. Because there's a fixed blue anchor on either side, the cord gets redirected there, and the pull starts widening the QRS complex.

That same pull also flattens the P wave, since there are two anchors here as well that redirect the tension.

Eventually, the QRS complex becomes so wide that it touches the T wave and merges with it.

## Nice to Know – Botched Blood Draw

When potassium levels are elevated, you should always keep pseudohyperkalemia in mind, especially if there are no matching symptoms or ECG changes.

It often results from the destruction of red blood cells (hemolysis) triggered during the blood draw. Common causes include prolonged tourniquet use, aspiration through narrow-bore needles, excessive muscle activity such as fist clenching, or improper handling of the samples afterward.

In these cases, potassium leaks out of the blood cells and distorts the values. A repeat blood draw performed correctly will usually show potassium levels back in the normal range.

## Hyperkalemia on the ECG

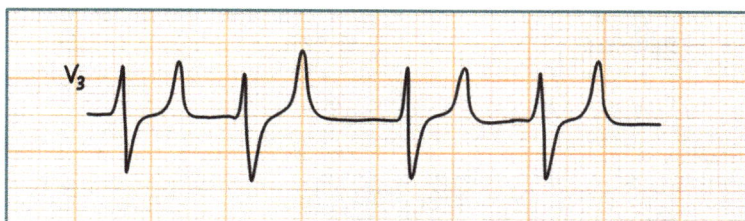

Here you can clearly see the excessively tall, sharply peaked T waves. If you count the boxes taken up by the QRS complex, you'll notice that the QRS duration is clearly prolonged. The P waves are so flat they're barely visible.

**Caution:** Don't confuse this with a De Winter T wave! In ischemia, you'd mainly see those in the precordial leads. Hyperkalemia, however, causes such tall T waves in all leads.

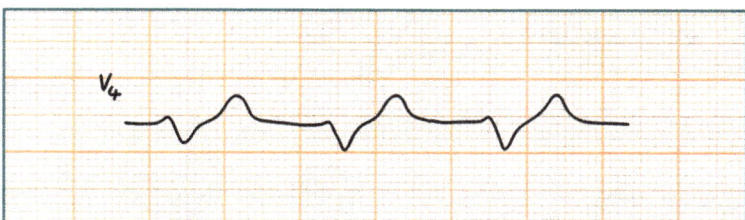

This example is even more dramatic. The T waves are less prominent here. Instead, you can see ST merging, which creates the typical sine wave pattern. As a result, the QRS complex is once again far too wide.

**Tip!**
When you think back to these mnemonics later, you might have trouble remembering which one goes with potassium and which with calcium.

**Try using the process of elimination:** since the screw clamp looks like a "C," it has to be the mnemonic for calcium. That means the cord analogy belongs to potassium.

## Hypokalemia

- T waves flattened

- Development of U waves

- ST depressions

Hypokalemia affects the heart rhythm in the exact opposite way. **It increases the excitability of pacemaker cells** by lowering the threshold potential, which raises the likelihood of malignant arrhythmias.

It starts with flattening of the T waves, which in severe cases may even turn negative. In addition, a so-called U wave can appear after the T waves. The whole thing looks a bit like a camel's hump.

These U waves can vary in size. Often they're small and subtle, but in pronounced hypokalemia they may even be taller than the preceding T wave. Along with these signs, ST depression is often present as well.

## An Easy Way to Remember the Effects of Hypokalemia

< That's Paul. Say hi!

The sagging backside and the humps show what happens in hypokalemia.

Let me introduce you to Paul. He's a camel, and as you might have noticed, he's got ridiculously short legs. But what he lacks in speed, he makes up with huge guts, which is why no predator dares to mess with him.

Here you can see what a U wave might look like. Notice how the T wave is pretty flat. You can tell because it barely rises higher than the P wave.

Keep in mind that the T and U waves can sit closer together than shown here. In that case, it may look like a T wave with two peaks rather than two separate waves. Also, the U wave doesn't necessarily have to be present. Hypokalemia can show up simply as very flat T waves and ST depression.

And that last part actually makes sense too, because with Paul's sagging backside, he can't keep the ST segment lifted, so it sinks a bit lower. But his long neck is enough to bring the line back up again after the T wave.

**Important!**
If the QT interval is extremely prolonged for any reason, the T wave can get so close to the next P wave that you might mistake it for a U wave.

Don't mix them up!

**Tip!**
Short on potassium? Think of 'short Paul' – same first letter, easy to recall.

You won't ever have trouble remembering the effects of hypokalemia again.

## Hypokalemia on the ECG

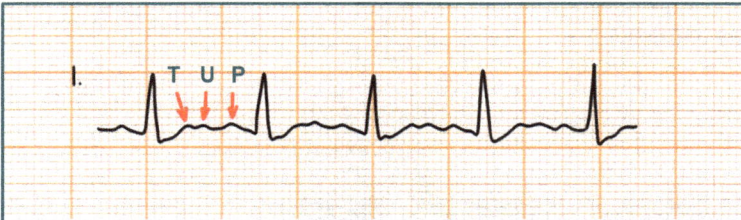

You can see that each T wave is followed by a U wave. Also, when compared to the P wave, it's clear that the T wave is too small. An upsloping ST depression is also present.

**Important:** Don't confuse this with atrial fibrillation! At first glance it might look a bit like it, since everything appears somewhat chaotic. But the key difference is that there's no arrhythmia of the ventricular complexes here.

# EVERYTHING AT A GLANCE

## Calcium
**Hypercalcemia**
- QT interval shortened

**Hypocalcemia**
- QT interval prolonged

Remember the mnemonic with the screw clamp and the rubber block! The two conditions behave in exactly the opposite way.

## Potassium
Can trigger severe arrhythmias!

**Hyperkalemia**
- T waves tall and peaked
- P wave flattening (up to disappearance)
- QRS duration prolonged
- Sine wave–like cardiac activity (ST merging)

The ECG changes appear in this order, depending on the severity of hyperkalemia. Picture the potassium hand pulling on the T-wave cord.

**Hypokalemia**
- T waves flattened or even negative
- Development of U waves
- ST depression

Short on potassium? Just remember short Paul.

As a little goodie at the end, here's a collection of ECG abnormalities that are all important in their own right but didn't quite get a chapter of their own.

They can all be explained fairly quickly and concisely, which is why they're grouped together here.

## Extrasystoles

This term refers to ventricular complexes that disrupt the regular rhythm by cheekily sneaking in between the normal heartbeats.

You can spot them because they show up earlier than you'd expect the next complex to appear. We distinguish between two types of extrasystoles, depending on where they originate.

- Supraventricular
- Ventricular

**Supraventricular extrasystoles** originate above the ventricles. That means the faulty impulses come either from the atria themselves or from the AV charging station.

You can recognize them by the P wave in front of the ventricular complex, which is often deformed or inverted. Typically, the QRS complex looks very similar to that of a normal heartbeat and is about the same narrow width.

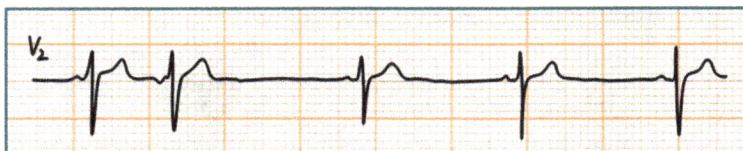

**Supraventricular Extrasystole:** This ECG was scaled down a bit so that a longer segment would fit into the displayed area. So don't bother counting boxes here!

The second cardiac cycle visible on the ECG shows the extrasystole with a negative P wave. Because of the premature ventricular depolarization, the next normal heartbeat can't occur, since the conduction system isn't ready yet. As a result, a longer pause follows, which then becomes visible on the ECG.

Ventricular extrasystoles originate directly from the ventricles. Unlike before, **there's no P wave here**, which makes sense, since the impulse doesn't come from the atria in the usual way but directly from the ventricles.

As a result, the **QRS complex is usually wide and deformed**. Makes sense too - an irregular depolarization of the ventricles also produces an irregular ventricular complex.

**Important!**
An extrasystole can appear either occasionally or in a regular pattern. If one occurs after every normal heartbeat, it's called a **bigeminy.**

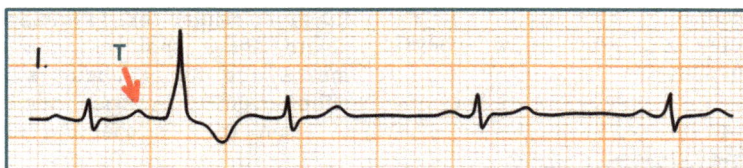

**Ventricular Extrasystole:** This ECG has also been scaled down for better overview. You can clearly see the unusually wide QRS complex, followed by an inverted T wave. It is huge compared to the others. The P wave is notably missing. The structure under the red arrow is actually the T wave of the previous complex, don't get them mixed up!

Interestingly, the distance between two normal complexes doesn't change here. That's purely by chance. The extrasystole just happened to fall in exactly the right spot, so the following heartbeat wasn't affected. In more technical terms, this is called an interpolated ventricular extrasystole. But a ventricular extrasystole can just as well result in a prolonged pause afterward.

## Nice to Know – Can You Feel It?

Occasional extrasystoles are usually harmless in people with a healthy heart. While some may notice them as palpitations ("heart skips" or "fluttering"), they typically don't cause any other symptoms.

When they occur only sporadically, they generally have no negative impact on physical performance.

## Tachycardias

As you probably already know, this term refers to the general case of the heart beating very fast (>100 bpm).

Physiological tachycardias occur as a normal response to physical exertion, stress, or fever. In clinical ECG interpretation, however, we're usually talking about pathological tachycardias.

Typically, these look like tachycardias without an obvious explanation. For example, a patient may be lying calmly on the bed yet still have a heart rate of 150.

Here, too, we distinguish between two types depending on their origin. Just like with extrasystoles, we talk about supraventricular tachycardias (SVT) and ventricular tachycardias (VT), depending on where they arise.

A **supraventricular tachycardia** is the somewhat less dangerous of the two. It can, for example, be triggered by a re-entry circuit when an accessory pathway bypasses the "AV station."

**The hallmarks are continuously present P waves and narrow, regularly shaped QRS complexes with each heartbeat.** This rules out a ventricular tachycardia.

That said, there are rare cases where wide, deformed complexes can appear. For that reason, our most important clue remains the presence of the P wave.

**Important!**
**Bad news:** In rare cases, P waves may be absent. Often because they're either hidden within broad QRS complexes or swallowed up by the closely following T wave.

**But that's not the norm.**

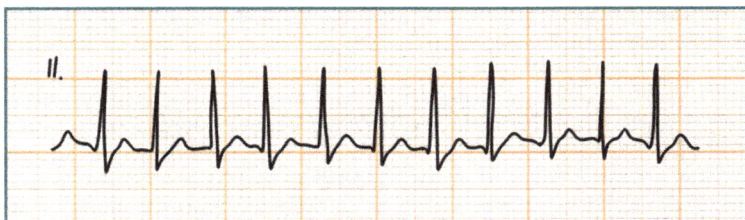

**Supraventricular tachycardia** with a rate of about 200 beats per minute. You have to look very closely to spot the occasional P waves, since they're quite small.

However, the regularly shaped QRS complexes also point to a supraventricular origin.

**Ventricular tachycardias** are much more dangerous, since at very high rates they can cause hemodynamic instability and, in the worst case, progress to ventricular fibrillation. Depending on rate and duration, a VT can lead to sudden cardiac death.

**That's why this is always an emergency!**

As with ventricular extrasystoles, **the P wave is usually absent before each complex. A wide, deformed QRS complex is typical,** since the tachycardic impulse originates directly from the ventricles.

VTs can appear either monomorphic (all complexes look alike) or polymorphic, such as in torsades de pointes (see below).

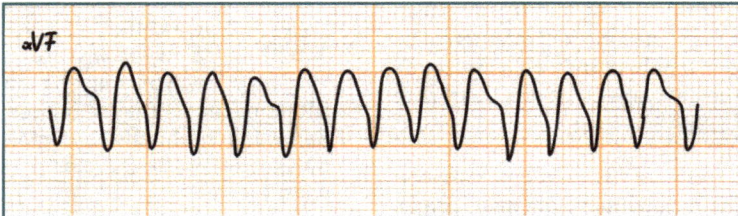

**Monomorphic Ventricular Tachycardia:** The wide, deformed complexes are clearly visible; no P waves can be seen anymore. These impulses are clearly originating from the ventricles.

The rate is even higher than in the previous example. Judging by the spacing between the complexes, this heart is beating about 250–300 times per minute.

You can probably imagine that at such a speed the ventricles can no longer fill properly.

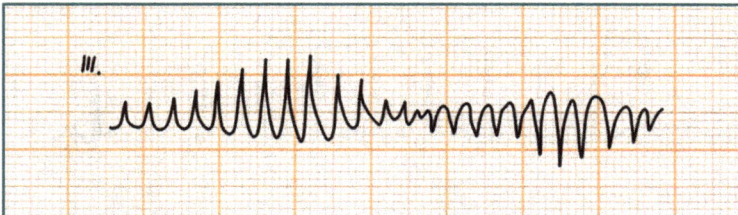

**Torsades de pointes:** A polymorphic ventricular tachycardia with a twisting QRS pattern. The complexes change their shape and amplitude, appearing to spiral around the baseline.

## Nice to Know – SVT vs. VT in Clinical Practice

The symptoms of a supraventricular tachycardia (SVT) range from being completely asymptomatic to palpitations, reduced performance, or occasional syncope. They are often self-limiting or can be terminated by vagal maneuvers, such as carotid massage.

If that doesn't work, medication with adenosine can help. It temporarily blocks the AV node for a few seconds, which can interrupt AV node–dependent tachycardias.

The ventricular form (VT) is initially treated with antiarrhythmics such as amiodarone, as long as there is some degree of hemodynamic stability. In the case of pulseless ventricular tachycardia (pVT), which no longer maintains circulation, immediate defibrillation is required.

The possible causes of ventricular tachycardias are varied. Most commonly they are due to coronary artery disease and myocardial infarction, but cardiomyopathies, myocarditis, and electrolyte disturbances also play a role.

## Ventricular Fibrillation

As already mentioned, a ventricular tachycardia can progress to ventricular fibrillation if nothing is done quickly. But there are many other possible causes as well, most notably ischemia (e.g., from a myocardial infarction).

Fibrillation is always a kind of last cry for help from the heart before no electrical activity remains.

Ventricular depolarization completely loses its regularity. All you see is chaotic scribbling, but no recognizable pattern.

The impulses vary wildly in size and shape, and the muscle contracts just as erratically. The heart twitches like crazy, and effective pumping is no longer possible.

**Immediate defibrillation is required!**

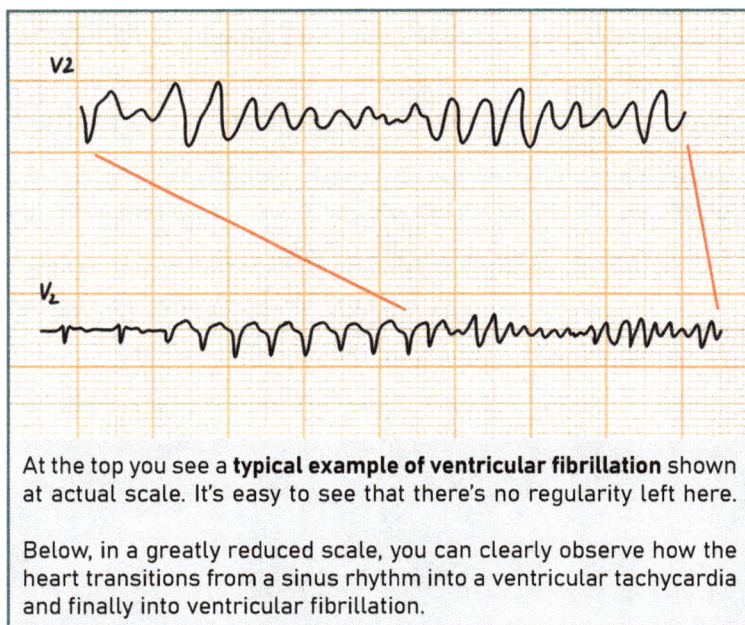

At the top you see a **typical example of ventricular fibrillation** shown at actual scale. It's easy to see that there's no regularity left here.

Below, in a greatly reduced scale, you can clearly observe how the heart transitions from a sinus rhythm into a ventricular tachycardia and finally into ventricular fibrillation.

# EVERYTHING AT A GLANCE

## Extrasystoles
**Supraventricular**
- P wave present (main clue)
- QRS complex usually narrow (rarely wide and deformed)

**Ventricular**
- No P wave (main clue)
- QRS complex wide and deformed

## Tachycardias
**Supraventricular**
- P waves present (main clue)
- QRS complexes usually narrow (rarely wide and deformed)

**Ventricular**
- No P waves visible (main clue)
- QRS complexes wide and deformed

Tachycardias and extrasystoles are easy to remember together, since they share the same distinguishing features.

## Ventricular Fibrillation
Ventricular tachycardias can progress into ventricular fibrillation, but it can also arise from other causes (e.g., ischemia)

It consists of chaotic, disorganized impulses of varying shape and amplitude (see image on the left). **The ventricles no longer contract, and immediate defibrillation is required!**

# THE TRIAL BY FIRE!
## PRACTICE EXAMPLES

### Time to Put Your Knowledge to the Test!

Here you'll find a small collection of ECGs to practice identifying pathologies. Grab a sheet of paper or open a text editor and try to write a short report for each example, just like you would in the clinic. If the QRS axis or the transition zone can't be assessed because of missing leads, simply skip that step.

Make sure to go through it systematically and don't miss any points. For reference, take another quick look at the summary on interpretation.

You'll find the solution to each example on the last page. Each ECG comes with a short patient history to give you a more realistic scenario.

You don't need to take the measurements yourself, they've already been done for you. If, for some reason, the computer couldn't automatically measure a value, you'll see "n.m." (not measurable) next to it. If you get stuck, just turn the book around and read the hint provided for each example.

In many of these exercises not all leads are shown, since it's tricky to fit them in original size on a small pocketbook page. But care was taken to ensure that the pathologies are still clearly diagnosable with the leads provided.

Some ECGs are recorded at a paper speed of 50 mm/s, so you can practice handling those as well. The relevant examples are flagged accordingly.

When all twelve leads need to be shown, or when a longer strip is necessary, the ECGs are scaled down. In those cases, you'll see the following symbol next to them:

**Have fun!**

## Example 1

**Tip!**
Looks like the conduction system's been struggling since the inflammation... where is the charge heading?

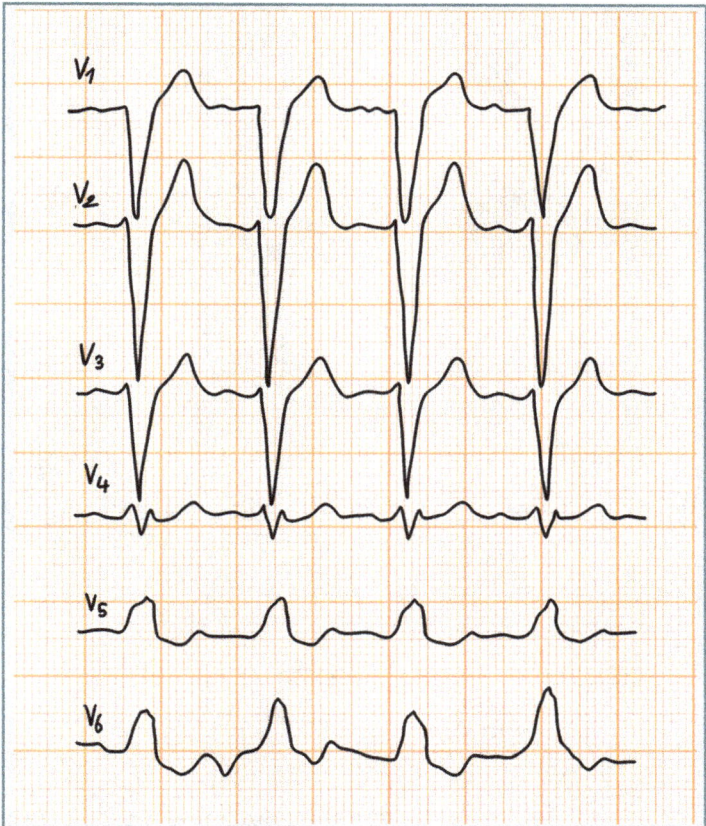

A 63-year-old patient presents to the outpatient clinic. She complains of general weakness, especially noticing limitations when climbing stairs. She reports having had a "heart inflammation" about three months ago and hasn't regained her previous level of fitness since.

| Rate = 90/min | P duration = 81 ms |
|---|---|
| PR interval = 190 ms | QRS duration = 195 ms |

## Example 2

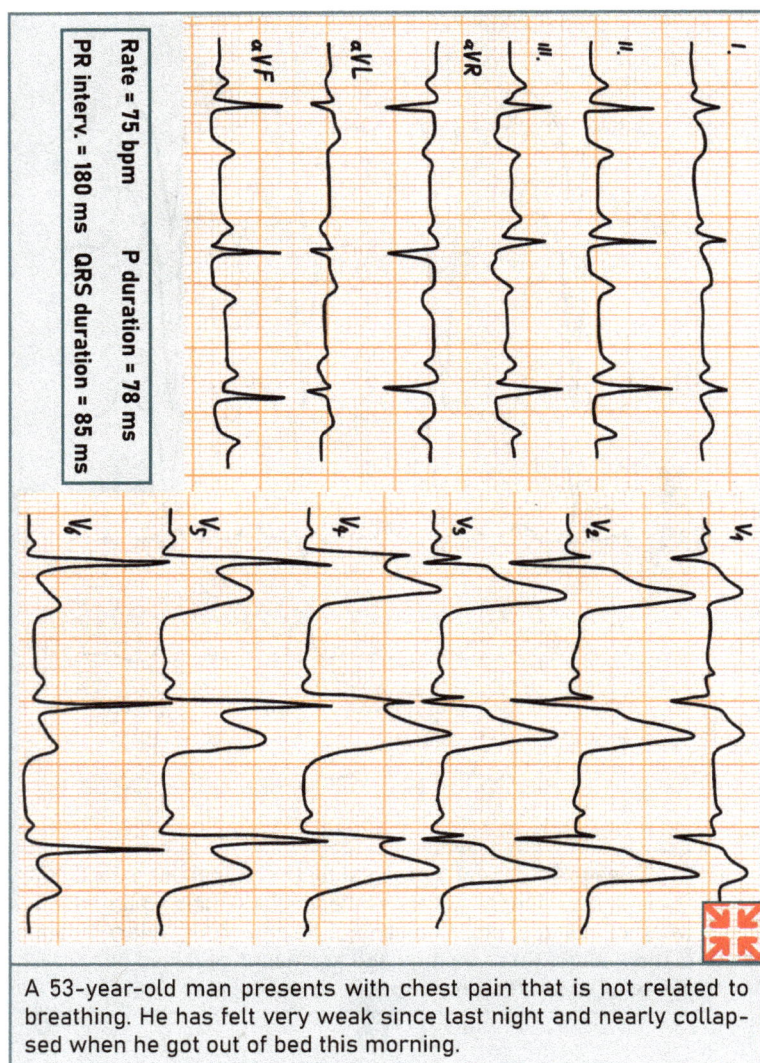

Rate = 75 bpm    P duration = 78 ms

PR interv. = 180 ms    QRS duration = 85 ms

A 53-year-old man presents with chest pain that is not related to breathing. He has felt very weak since last night and nearly collapsed when he got out of bed this morning.

Tip!
What ST-segment changes do you see?
What do they suggest?

## Example 3

Tip!
This ECG looks almost completely fine.
Take a closer look at the values.

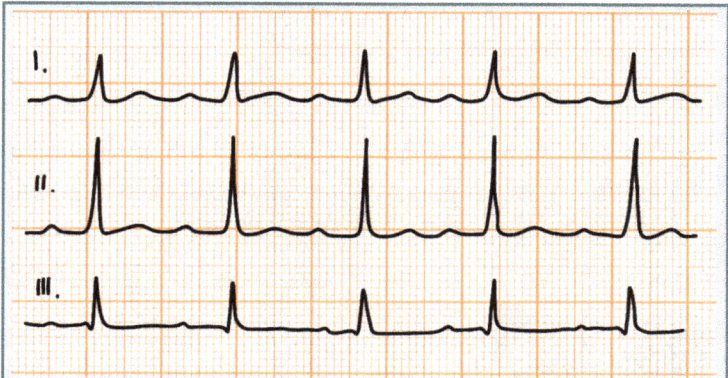

I.

II.

III.

An 18-year-old patient is having her first annual check-up. She says she feels fine and healthy. She doesn't smoke, rarely drinks, and exercises regularly. The examination itself is unremarkable, but in the ECG you notice the following...

| | |
|---|---|
| Rate = 83 bpm | P duration = 81 ms |
| PR interval = 230 ms | QRS duration = 79 ms |

## Example 4

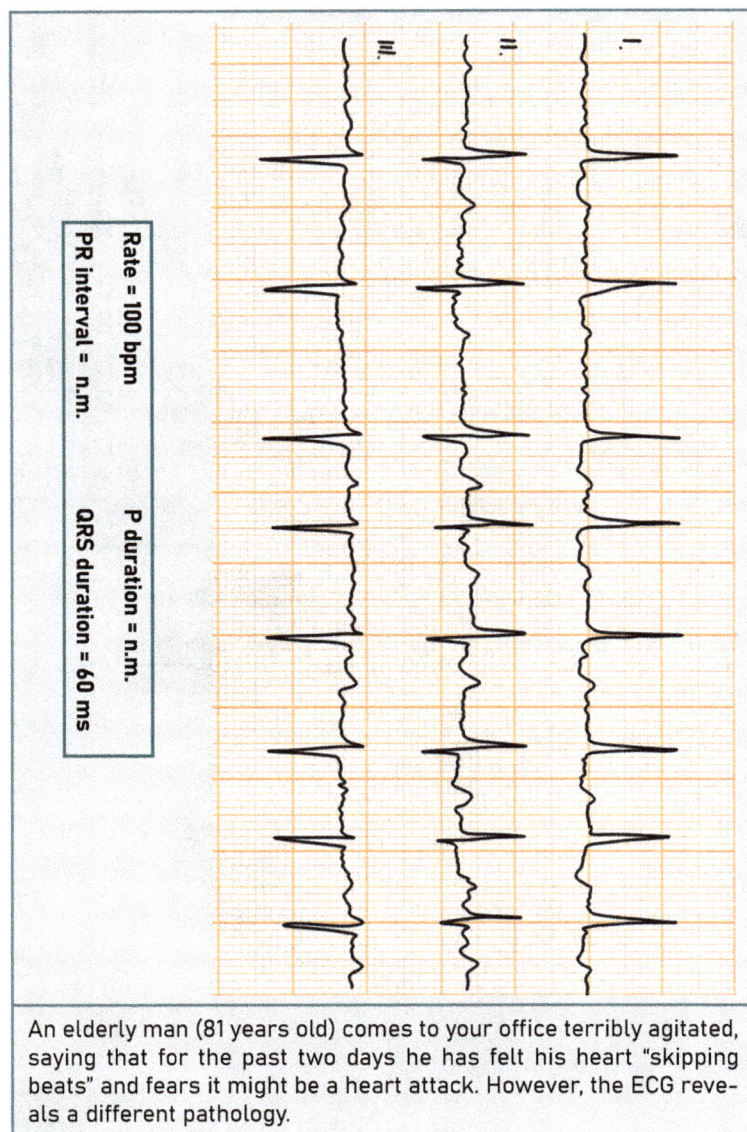

Rate = 100 bpm

PR interval = n.m.

P duration = n.m.

QRS duration = 60 ms

An elderly man (81 years old) comes to your office terribly agitated, saying that for the past two days he has felt his heart "skipping beats" and fears it might be a heart attack. However, the ECG reveals a different pathology.

**Tip!**

Have you noticed that the ventricular complexes appear irregular? What could be causing that?

## Example 5

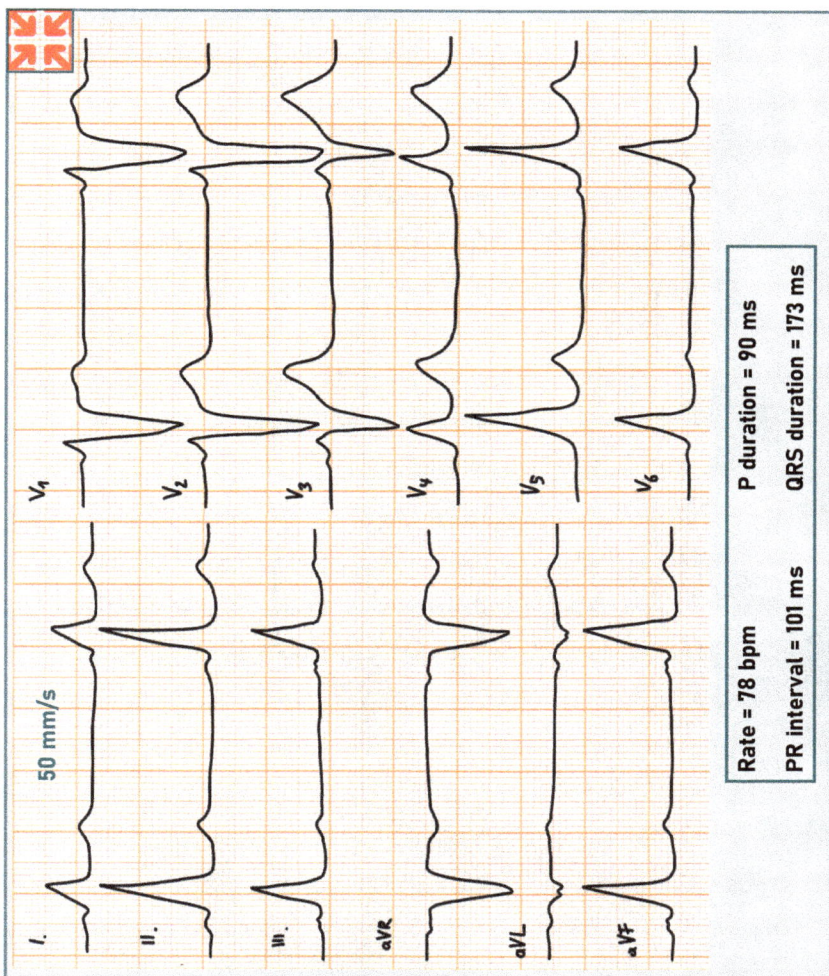

50 mm/s

I.

II.

III.

aVR

aVL

aVF

V₁

V₂

V₃

V₄

V₅

V₆

Rate = 78 bpm

PR interval = 101 ms

P duration = 90 ms

QRS duration = 173 ms

A 36-year-old patient tells you that she has been experiencing episodes of tachycardia for about three months. These often occur shortly after physical exertion, such as climbing stairs. The episodes last around 10–15 minutes, during which she feels pronounced palpitations and breaks out in a sweat. At the time of the examination, however, no tachycardia is present, and she feels "normal."

**Tip!**
So why has the P wave moved so close to the QRS complex here?

## Example 6

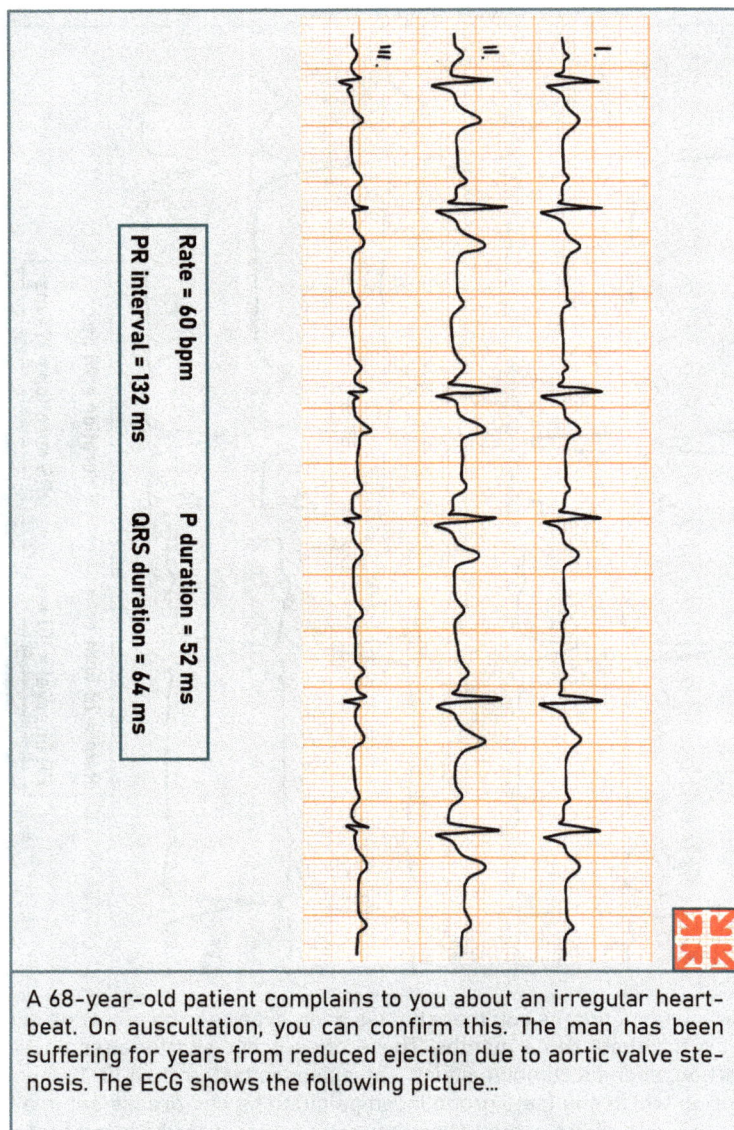

Rate = 60 bpm    P duration = 52 ms

PR interval = 132 ms    QRS duration = 64 ms

A 68-year-old patient complains to you about an irregular heart-beat. On auscultation, you can confirm this. The man has been suffering for years from reduced ejection due to aortic valve stenosis. The ECG shows the following picture...

**Tip!**
What's that little bump that sometimes shows up between two heartbeats? Could it be a clue?

## Example 7

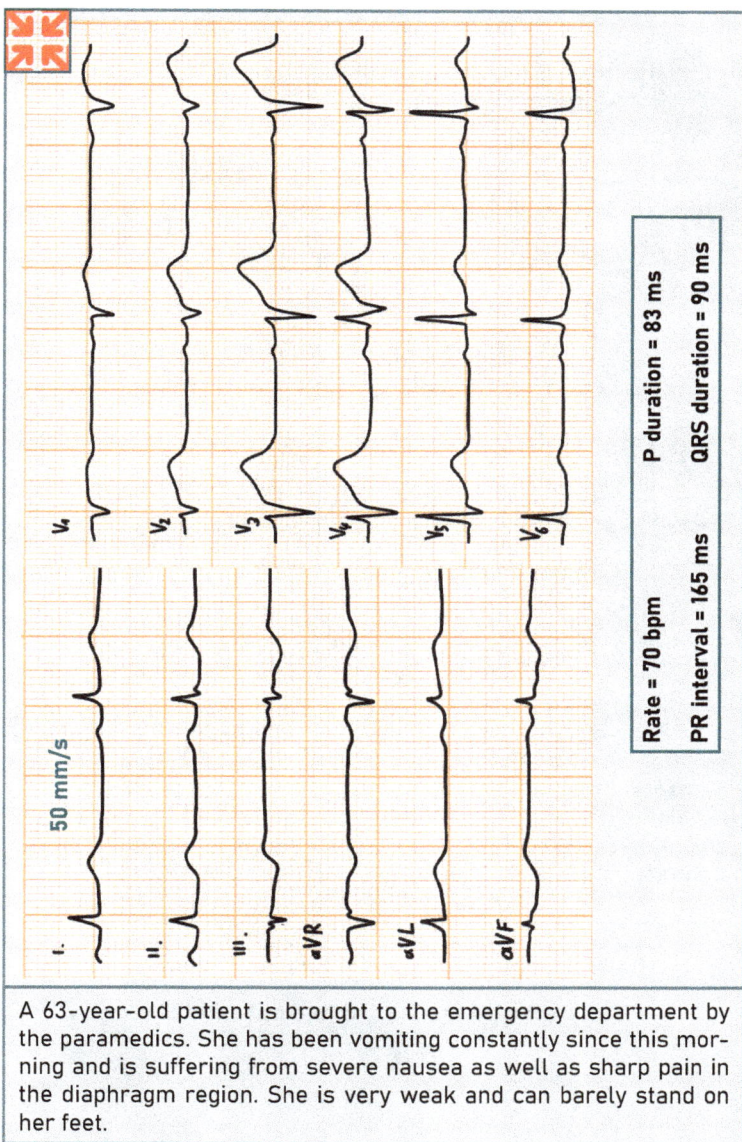

**Rate = 70 bpm**

**P duration = 83 ms**

**PR interval = 165 ms**

**QRS duration = 90 ms**

50 mm/s

A 63-year-old patient is brought to the emergency department by the paramedics. She has been vomiting constantly since this morning and is suffering from severe nausea as well as sharp pain in the diaphragm region. She is very weak and can barely stand on her feet.

**Tip!**
You might find a clue here in some of the precordial leads. Picture the isoelectric line in your mind.

## Example 8

**Tip!**
The right ventricle was put under severe strain. Could something have been damaged as a result?

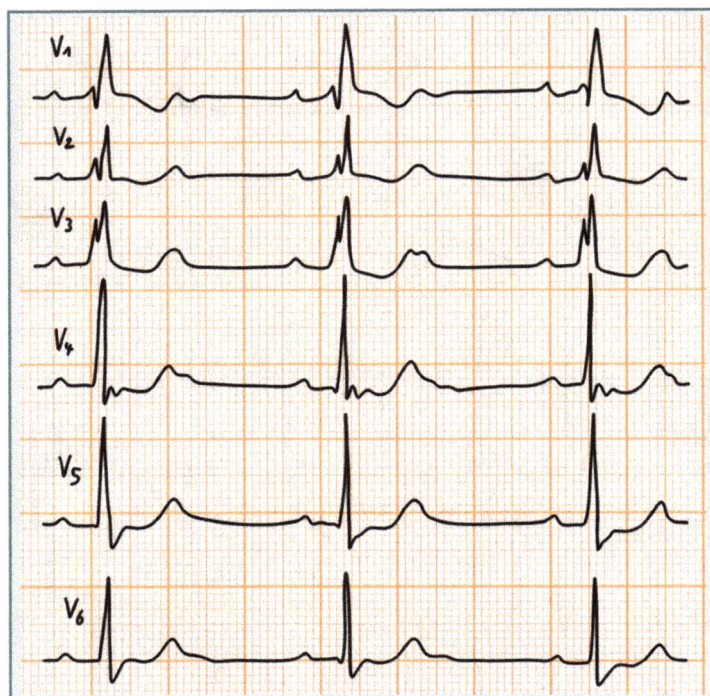

A 45-year-old, very athletic man was admitted to the hospital after returning from a business trip to Japan. It turned out he had suffered a massive pulmonary embolism caused by a thrombus that had broken loose from the leg veins. Echocardiography revealed an acute right heart strain as a result. After treatment of the embolism, an ECG was recorded.

| Rate = 55 bpm | P duration = 82 ms |
|---|---|
| PR interval = 195 ms | QRS duration = 125 ms |

## Example 9

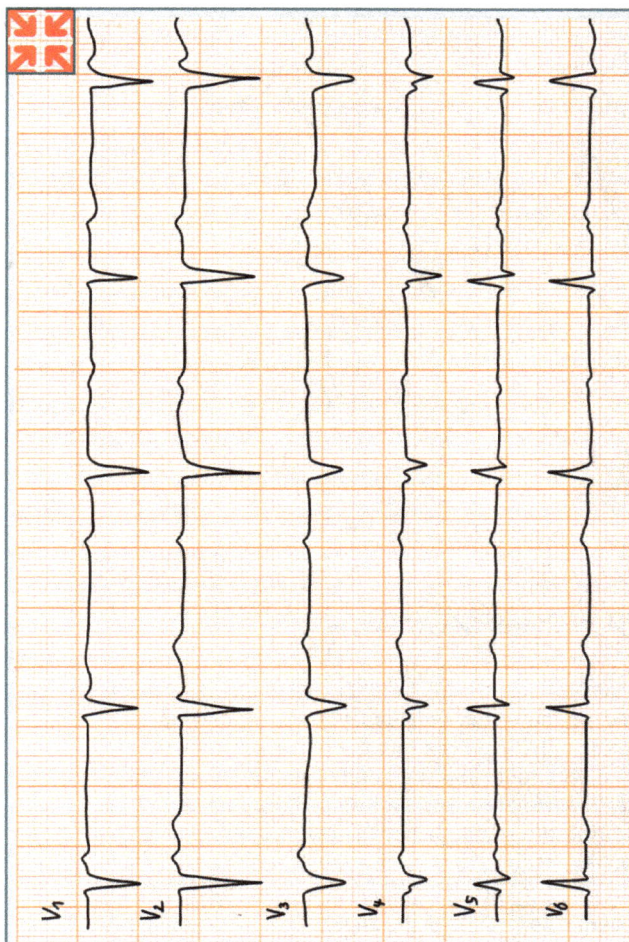

Rate = 46 bpm

PR interval = n.m.

P duration = 65 ms

QRS duration = 95 ms

During your night shift, you're called in because a 73-year-old patient has been brought in by the paramedics. She collapsed several times during the night when trying to get to the bathroom. Her medical records reveal that she had already been treated two years ago for a second-degree AV block, Mobitz II. The notes also show that back then, despite thorough counseling about the risks, she refused pacemaker implantation.

Tip!
Is this really still a second-degree AV block? At first glance it may seem so, but something has changed.

## Example 10

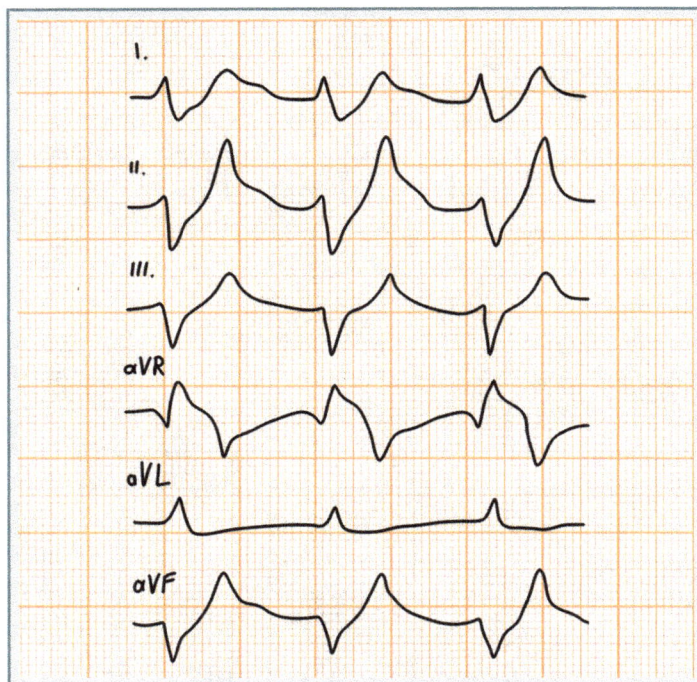

A 75-year-old man is admitted because of nausea, dizziness, and shortness of breath. From the initial interview it becomes clear that he is an alcoholic, and the condition of his liver and kidneys is borderline. He recounts that his primary care physician told him some time ago: "If your kidneys don't recover, you'll soon need dialysis."

**Tip!**
So what exactly are the kidneys responsible for? What builds up in the blood if they stop doing their job properly? And what could that mean in relation to the ECG?

| Rate = 72 bpm | P duration = n.m. |
|---|---|
| PR interval = n.m. | QRS duration = 192 ms |

## Example 11

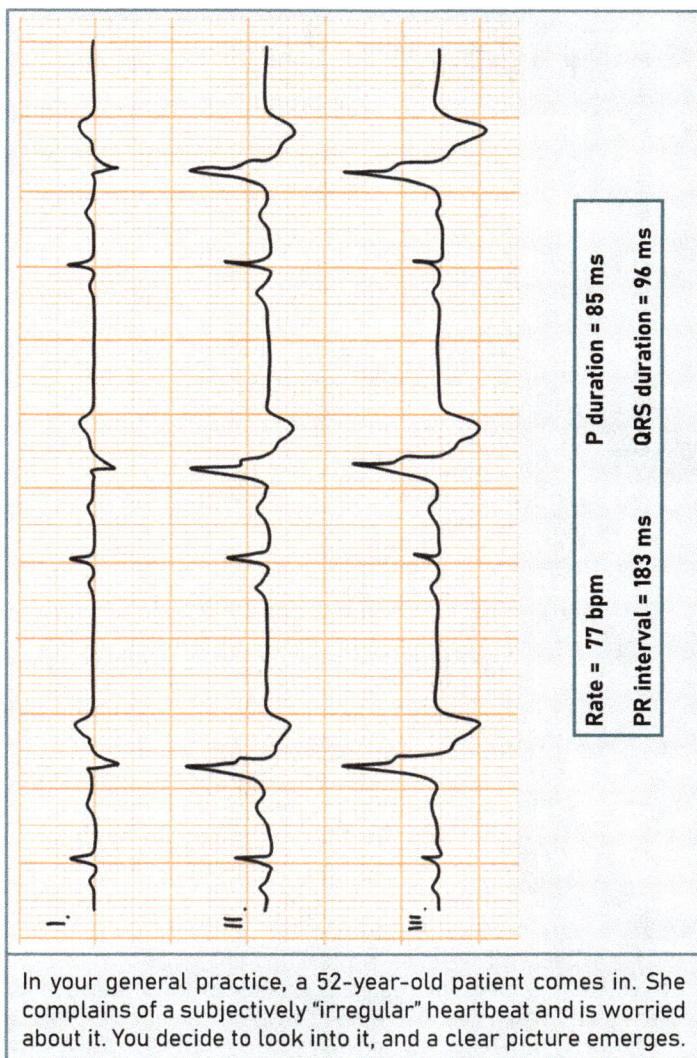

Rate = 77 bpm     P duration = 85 ms

PR interval = 183 ms     QRS duration = 96 ms

In your general practice, a 52-year-old patient comes in. She complains of a subjectively "irregular" heartbeat and is worried about it. You decide to look into it, and a clear picture emerges.

**Tip!**

Why do the heartbeats here always appear in pairs? And why is the second one each time wide and misshapen? It seems to have a different origin than the first, normally shaped one…

## Example 12

Rate = 43 bpm    P dur. = 79 ms

PR interv. = 186 ms    QRS dur. = 102 ms

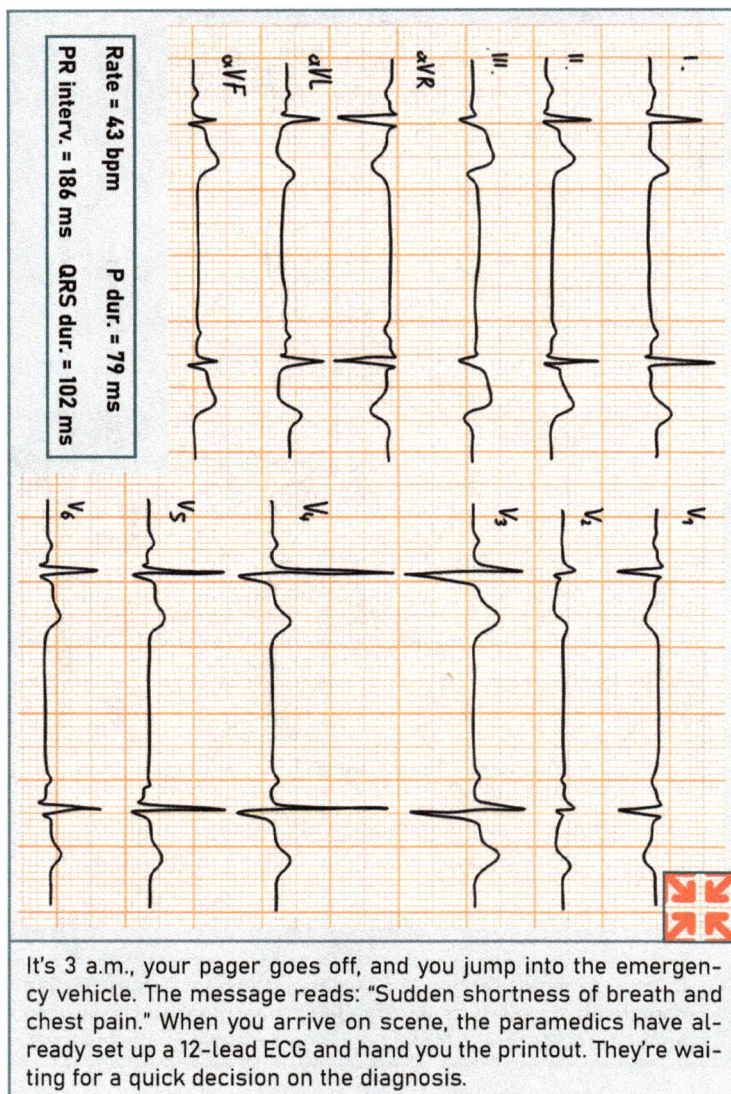

I
II
III
aVR
aVL
aVF
V1
V2
V3
V4
V5
V6

It's 3 a.m., your pager goes off, and you jump into the emergency vehicle. The message reads: "Sudden shortness of breath and chest pain." When you arrive on scene, the paramedics have already set up a 12-lead ECG and hand you the printout. They're waiting for a quick decision on the diagnosis.

**Tip!**
Which diagnoses could be considered with these symptoms that can actually be identified on the ECG? Pay close attention to the ST segments.

## Example 13

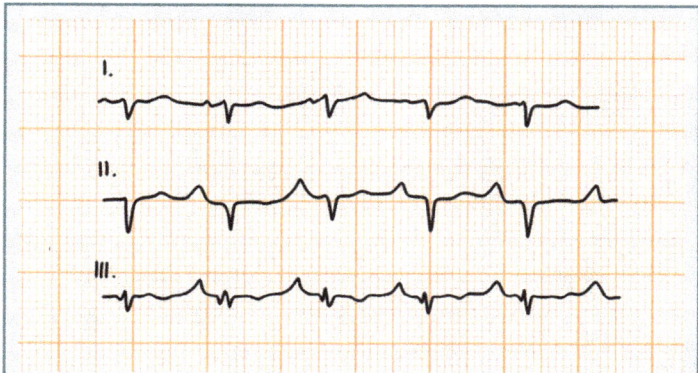

A man in his mid–60s comes in for his annual check-up. Though, to be fair, "annual" might be a bit of a stretch - your last encounter with him was five years ago. On his ECG, one small detail stands out...

| | |
|---|---|
| Rate = 110 bpm | P duration = 105 ms |
| PR interval = 168 ms | QRS duration = 84 ms |

**Tip!**

Why is the T wave appearing before the QRS complex? Or could it be that this tall, sharp deflection isn't a T wave at all? Here you need to be careful not to mix things up. Based on its position, this is much more likely to be the P wave...

## Example 14

**Rate = 80 bpm**

**PR interval = 142 ms**

**P duration = 79 ms**

**QRS duration = 105 ms**

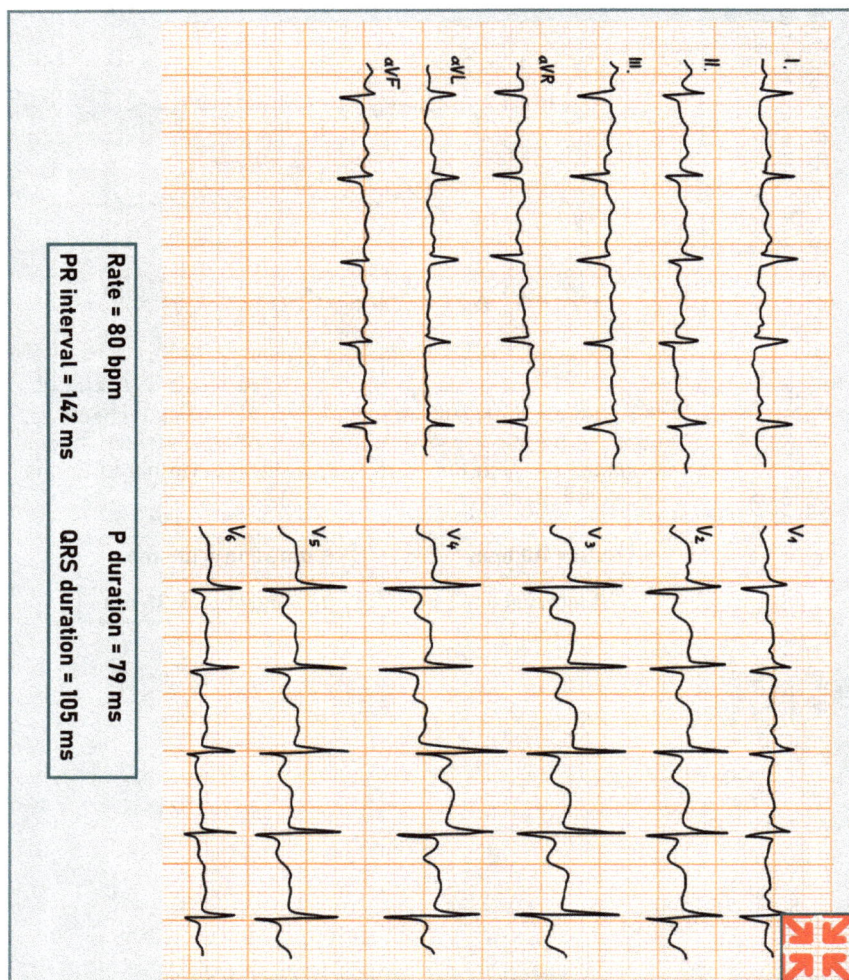

A 75-year-old woman comes to see you and reports that she often experiences dizzy spells accompanied by nausea during physical exertion. These episodes last about five minutes, forcing her to stop and rest. Her resting ECG appears normal, so you put her on the treadmill and monitor the ECG during exercise. What you observe is the following...

**Tip!**

As physical exertion increases, so does the oxygen demand of the heart muscle. Could this be pointing to a problem with oxygen supply?

## Übung 15

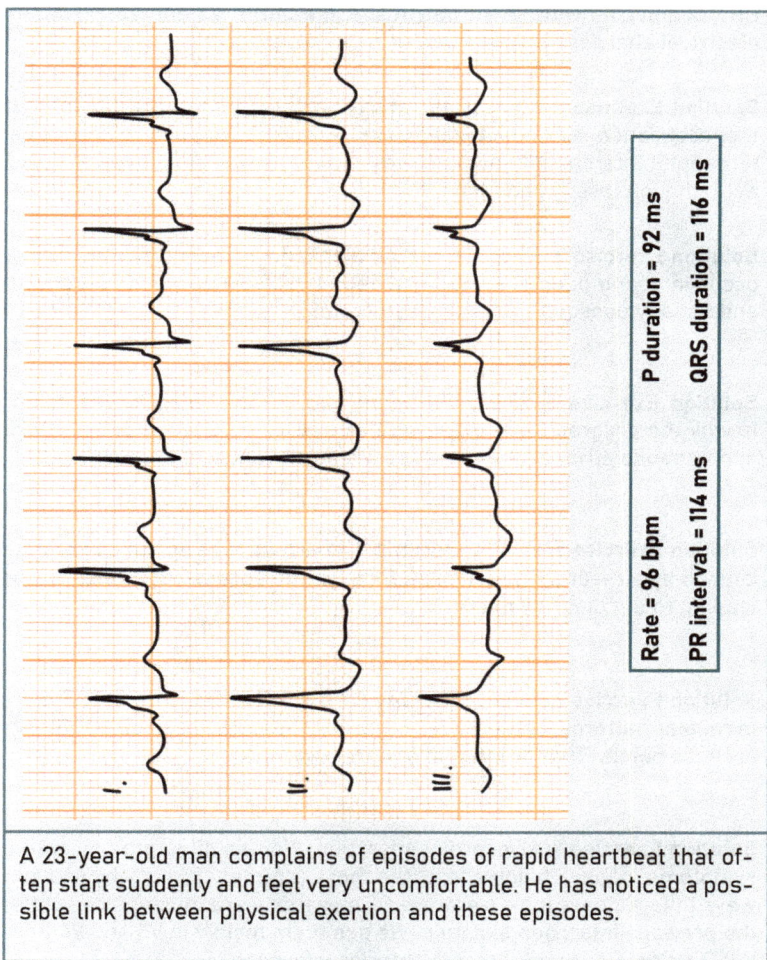

Rate = 96 bpm  P duration = 92 ms

PR interval = 114 ms  QRS duration = 116 ms

A 23-year-old man complains of episodes of rapid heartbeat that often start suddenly and feel very uncomfortable. He has noticed a possible link between physical exertion and these episodes.

**Tip!**
What looks suspicious about the values?
Take a close look at the QRS complex—the devil is in the details.

**Solution Exercise 1:** Left bundle branch block. You can see the deep S waves in V1 and V2 (and still in V3), as well as the positive, wide, unusually shaped QRS complexes with small notches at the top in V5 and V6. This means the electrical charge is taking a detour as it moves from right to left.

**Solution Exercise 2:** Myocardial infarction. Even better if you also figured out the localization. Marked elevations in V2 through V4 point to an extensive anteroseptal infarction. In addition, the lateral wall is also involved (V5, V6, and aVL), though less pronounced.

**Solution Exercise 3:** First-degree AV block. An incidental finding that can also occur in young people. In well-trained athletes, it can even be physiological and usually does not cause any symptoms.

**Solution Exercise 4:** Atrial fibrillation. No P waves can be identified, which is why the program doesn't provide values for P duration or PR interval. The spontaneous atrial activity leads to irregular ventricular complexes.

**Solution Exercise 5:** WPW syndrome. The tachycardia episodes are most likely caused by a re-entry mechanism through the accessory conduction pathway, leading to circulating impulses.

**Solution Exercise 6:** Second-degree AV block, Mobitz II. Conduction drops out in regular patterns. In this case, every third beat. You can see a solitary P wave at those points. The PR interval is unremarkable.

**Solution Exercise 7:** Myocardial infarction. Women often present with atypical symptoms, nausea being one of the most common. That's why you should always take a closer look. Here, the ST elevations are much more subtle than in the previous infarction example. We see them mainly in V3 and V4 (minimal in V2). This points primarily to an anterior infarction.

**Solution Exercise 8:** Right bundle branch block. The pulmonary embolism put the right ventricle under prolonged strain, damaging the right bundle branch. We see M-shaped complexes in V1, V2, and even V3. The persistent S wave in V5 and V6 is clear. The QRS duration is prolonged.

**Solution Exercise 9:** Third-degree AV block. The second-degree block has progressed to a complete block. Since this exercise is fairly challenging, here's the ECG broken down again with a detailed explanation:

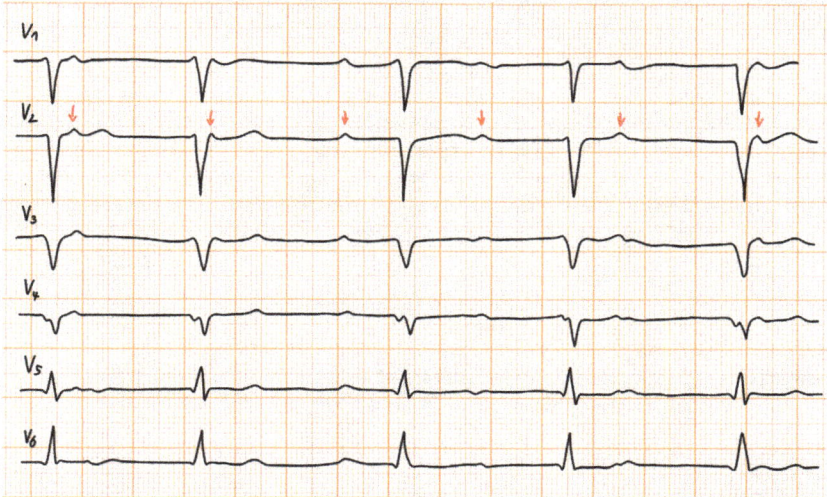

Typical of third-degree AV block is that the P waves (arrows) can no longer be clearly assigned to the QRS complexes; they are scattered all over. Some appear right in the middle of a complex, others just after the T wave. In addition, there are more P waves than QRS complexes, which is another key clue. Ventricular activation is taken over by a slower escape rhythm.

**Solution Exercise 10:** Hyperkalemia. Due to acute renal failure, the patient's body is no longer able to excrete electrolytes adequately. As a result, potassium and other minerals, along with various uremic substances, accumulate in the blood. Especially in leads III and aVF, we can already see sinus wave–like heart actions.

**Solution Exercise 11:** Ventricular extrasystoles. Wide and tall ventricular complexes point to an origin in the ventricles. Since an extrasystole occurs after each heartbeat, this is called bigeminy. As long as there is no impairment of cardiac function, it is harmless.

**Solution Exercise 12:** Myocardial infarction. ST elevations are visible in leads II, III, and aVF, indicating involvement of the inferior wall. As a small bonus, you might also have noticed the M-shaped complex in V1 and the persistent S wave, but what argues against a right bundle branch block is that the QRS duration is normal.

**Solution Exercise 13:** P pulmonale. This points to a possibly overloaded right atrium due to increased resistance in the pulmonary circulation. One possible cause could be pulmonary valve stenosis. This patient needs an echocardiography for further evaluation. What you should also have noticed is the significant rightward deviation of the QRS axis, an important detail that must not be missing from your report.

**Solution Exercise 14:** Anteroseptal ischemia with lateral involvement. Admittedly, a somewhat more challenging example. In the stress ECG you can see downsloping ST depressions in V2 through V4, which indicate oxygen deficiency of the septum and the anterior wall. In addition, the lateral wall is also affected in V5 and V6, though less strongly. Since this only occurs under stress, it is not yet a myocardial infarction, but early action should already be taken here.

**Solution Exercise 15:** WPW syndrome. In leads I and II we see distinct delta waves blending into the QRS complexes. In lead III, it almost looks like an extra wave within the complex. The PR interval is shortened.

# How was it?

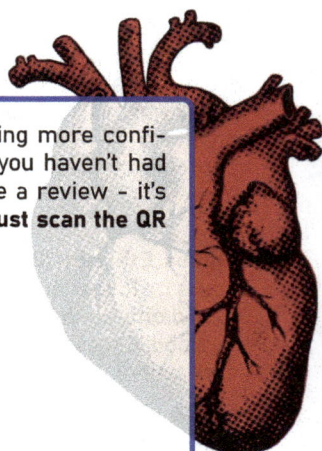

Whether you're now an ECG pro or just feeling more confident, I'd love to hear your final thoughts. If you haven't had a chance yet, please take a moment to leave a review - it's the greatest support a creator can ask for. **Just scan the QR Code below.**

**Thank you so much!**

**Philipp**

## Closing Words

**Thanks a lot for choosing this book!** Hopefully, you now feel a lot more confident when someone hands you an ECG strip and can approach the task with greater assurance.

If you feel like you haven't figured everything out yet, or if you still struggled with the practice examples, don't worry. A complex subject like ECGs takes patience and practice. Stick with it and don't get discouraged.

During your clinical rotations and beyond, try to take every opportunity to interpret an ECG. Even if someone else has already done the interpretation, go through it again yourself. And if at some point you need a refresher, this book will help you get back on track quickly - "at a glance." Hopefully, it will serve you well for a long time to come.

Remember, very rarely does everything depend on the ECG alone, so no need to panic. In most cases you can rely on other tests for support, such as blood work, echocardiography, or other imaging. Most importantly, think about the patient's history to always keep the big picture in mind.

**And don't forget, at the end of the day, nobody ever became a pro overnight!**

All the best for the future!

**Imprint**

**Publisher and Rights Holder:**

Philipp Bergher
c/o COCENTER
Koppoldstr. 1
86551 Aichach
Germany

This address is provided through an official service for legal correspondence. The rights holder is based in Innsbruck, Austria.

**kontakt@ekghandbuch.com**
This email address is intended solely for legal matters, business inquiries, or correction notes. Feedback and suggestions are welcome via Amazon reviews. Likewise, returns and refunds will only be handled directly via Amazon.

**Produced/Printed by**
**Amazon Kindle Direct Publishing:**

Amazon EU Media S.à.r.l
5 Rue Plaetis
L-2338
Luxembourg

**Printing location may differ.**
See final page.

**ISBN Publisher:**

Staten House - Founder Technology LLC
447 Broadway
2nd Floor
New York, NY 10013
USA

Date of Publication: 08/2025

www.ingramcontent.com/pod-product-compliance
Lightning Source LLC
Chambersburg PA
CBHW071154200326
41519CB00018B/5228